GIS

Short Introductions to Geography are highly accessible books designed to introduce key geographical concepts to students. Taking a concise approach, these introductions convey a sense of the intellectual liveliness, differing perspectives, and key debates that have developed around each concept. The central ground is covered and readers are encouraged to think in new and critical ways about concepts that are core to geographical study. The series will also serve a vital pedagogic function, encouraging students to recognize the significance and value of both conceptual and empirical analyses. Instructors meanwhile will be assured that students have an essential conceptual reference point.

Published

GIS: A Short Introduction
Nadine Schuurman

Place: A Short Introduction
Tim Cresswell

In Preparation

Maps: A Short Introduction
Matthew Edney

Territory: A Short Introduction
David Delaney

GIS

a short introduction

Nadine Schuurman

BLACKWELL PUBLISHING
350 Main Street, Malden, MA 02148-5020, USA
9600 Garsington Road, Oxford OX4 2DQ, UK
550 Swanston Street, Carlton, Victoria 3053, Australia

First published 2004 by Blackwell Publishing Ltd

5 2008

Library of Congress Cataloging-in-Publication Data

Schuurman, Nadine.
GIS : a short introduction / Nadine Schuurman.
p. cm. – (Short introductions to geography)
Includes bibliographical references and index.
978-0-631-23532-3 (alk. paper) — 978-0-631-23533-0 (alk. paper)
1. Geographic information systems. I. Title. II. Series.

G70.212.S38 2003
910′.285—dc21

2003005917

A catalogue record for this title is available from the British Library.

Set in 10/12 pt Palatino
by Kolam Information Services Pvt Ltd, Pondicherry, India

The publisher's policy is to use permanent paper from mills that operate a sustainable forestry policy, and which has been manufactured from pulp processed using acid-free and elementary chlorine-free practices. Furthermore, the publisher ensures that the text paper and cover board used have met acceptable environmental accreditation standards.

For further information on
Blackwell Publishing, visit our website:
www.blackwellpublishing.com

Contents

For Jen and Capes

Figures

Full color versions of all these figures can be seen at
http://www.blackwellpublishing.com/schuurman

Tables

Series Editors' Preface

Short Introductions to Geography are highly accessible books, written by leading scholars, that are designed to introduce key geographical ideas to students and other interested readers. Departing from a traditional subdisciplinary review, they seek to explain and explore central geographical and spatial concepts. These concise introductions convey a sense of the intellectual liveliness, differing perspectives, and key debates that have developed around each concept. Readers are also encouraged to think in new and critical ways about concepts that are core to geographical study. The series serves a vital pedagogic function of encouraging students to recognize how concepts and empirical analyses develop together and in relation to each other. Instructors meanwhile will be assured that students have an essential conceptual reference point, which they can supplement with their own examples and discussion. The short, modular format for the series allows instructors to combine two or more of these texts in a single class, or to use the text across classes with a distinctive subdisciplinary focus.

Nicholas Blomley

Geraldine Pratt

Author's Acknowledgments

The intellectual content of this book gestated over several years as I often tried to explain GIS as a *science* to human geographers, and describe it as a *social process* to GIScientists. In both cases, I was forced to examine and strengthen my arguments. This book is directed at both these realms of geography which I continue to straddle.

I have many people to thank for assisting and supporting me in bridging these two subdisciplines. Gerry Pratt and Nick Blomley, the series editors, provided encouragement and incentive to write this book. Tom Poiker offered several insights, and I am grateful for his intellectual support. Fang Chen, whose work is featured in Chapter 4, provided invaluable research material. Mike Hayes, the principal investigator of the population health research also featured in Chapter 4, provided an open, dynamic research environment to develop new ideas. My research collaborator, Suzana Dragicevic, is the inspiration behind the multi-criteria evaluation (MCE) aspect of our population health research, and Darrin Grund, our GIS systems analyst, provided invaluable assistance more than once. Sarah Elwood and Renée Sieber, two of my valued colleagues, provided ideas for this project. Our departmental cartographer, John Ng, is responsible for many of the illustrations, and I couldn't have embarked on this project without his help. Jasper Stoodley, our systems administrator is the backbone of everything we do in GIS at SFU, and deserves praise. Rob Fiedler, a GIS graduate student, provided help with a key example. Maggie Isenor made an important contribution to the discussion of data sharing. My editor at Blackwell, Debbie Seymour, has been perfect when the manuscript was not, and I thank her for attention to many details that surely would have escaped me. The idea of this book emerged during a lunch with Sarah Falkus, former Blackwell Geography editor, whose support I would like to acknowledge. Peter Baehr clarified those murky waters of epistemology; I thank him for his care in reading parts of the manuscript and for his enthusiasm about it. My mother and father have always been excited about ideas, and I would like to thank them for instilling that passion in me. Finally, I am indebted to my reviewers, Eric Sheppherd and Stacy Warren who are certainly responsible for strengthening the book. All mistakes, errors and shortcomings are, of course, my own.

Funding for research presented in this book was supplied in part by a Canadian Population Health Initiative (CPHI) collaborative research grant and a Social Sciences and Humanities Research Council (SSHRC) grant (#401-2001-1497).

1

Introducing the Identities of GIS

The Success of GIS

GIS is enjoying a boom. It is increasingly recognized by disciplines outside geography, and, to many, epitomizes "modern geography." Software sales exceed seven billion US dollars annually; students flock to GIS classes in colleges and universities; on-board navigation systems are the mark of a luxury car; police officers are routinely trained in GIS; organ donation has been rationalized using GIS; epidemiologists use GIS to identify clusters of infectious disease; archaeologists use it to map sites; and Starbucks® is reputed to use GIS to site its very successful coffee shops. Indeed, the list of GIS uses is extraordinarily comprehensive; the technology pervades many aspects of modern life. Technical advances in GIS have proceeded before our ability to realize and understand its potential effects. Means of integrating the pervasive role and influence of GIS have not kept pace with the development and proliferation of the technology. Indeed, many people do not recognize the acronym; they are even less likely to be able to tell you how GIS has affected their everyday lives. But it has.

This book is designed to inform the reader about precisely how GIS affects them as well as myriad social processes. It introduces what GIS is, how it is understood differently in different contexts, how it works, the importance of data, how data are stored and manipulated, and what contemporary GIS research looks like. It surpasses a mere descriptive account, however, in that it introduces and explores philosophical implications of using GIS whether for research, planning, marketing, environmental management, or other tasks. These are complicated issues, but necessary to understand the full scope of GIS. The book is unique in that it is not a "how-to" guide for technically minded students. Rather it aims to introduce the intellectual territory and practice of GIS to a wide variety of people. Indeed, its audience is catholic encompassing interested physical and social geographers, GIS users, students, and anyone who has worked with or wondered about GIS. It is designed to illustrate how GIS affects people by changing the way they do everyday tasks from wayfinding to data collection.

Given the ubiquity of GIS, its value would seem to be undisputed. Except perhaps in the discipline of geography. Academic geographers have a love/hate relationship with GIS perhaps because we are so close to its faults and biases. This relationship is made more complex because GIS represents only one lens on the physical and social world, but this is the face to which the world has had the most exposure to, especially in the last decade. Incoming undergraduate students routinely know of GIS, but are less likely to be familiar with qualitative research techniques used by some human geographers, or about the use of ground-penetrating radar by geomorphologists. The ubiquity of GIS has perhaps colored the perception of geography, and this has a bearing on the identity of all geographers.

It may be surprising then to learn that GIS does not have its own fixed and secure identity. It suffers from the scourge of being many things to many people. To a municipality, GIS is the software that allows planners to identify residential, industrial, and commercial zones. It maps the exact location and survey coordinates of each taxable property, and provides answers to queries such as: "how many properties would be affected by the addition of an extra lane to Highway 1 between 170 and 194th streets?" To a university researcher who must define the boundaries of communities that enjoy varying health outcomes, GIS is a different animal. It is not a piece of software, but a scientific approach to the problem: "how do we define crisp boundaries to demarcate fuzzy and changeable phenomena?" The latter is a fundamentally philosophical issue that must be resolved through computing. These two types of questions are very different. One is interested in "where" spatial entities are or might be while the other is concerned with "how" we encode spatial entities (e.g., communities, urban/rural areas, forests, roads, bridges, and anything that might appear on a map), and the repercussions of different methods of analysis on answers to geographical questions. Both are asked, however, with respect to GIS, and they point to the myriad ways that GIS can be defined and perceived – the basis of its identity problem. And identity, as a cursory review of present world politics will confirm, is closely linked to history.

Where Does GIS Come From? A Technical History

The roots of GIS' identity problem date back to the 1960s when the technology and epistemology that underlie it were first being developed. Methods of computerizing cartographic procedures were coincident with the realization that mapping could segue neatly into analysis. In 1962, Ian McHarg, a landscape architect introduced the method of "overlay" that was later to become the *sine qua non* methodology of GIS. He was searching for the optimal route for a new highway that would be associated with suburban development. His goal was to route the highway such that its path would involve the least disruption of other "layers" of the landscape including forest cover, pastoral valleys, and existing semirural housing. He took multiple pieces of tracing paper, one representing each layer, and laid them over each other on a light table. By visually examining their intersections, he was able to "see" the only logical route. The process of overlaying map layers is depicted in Figure 1.1. Ironically,

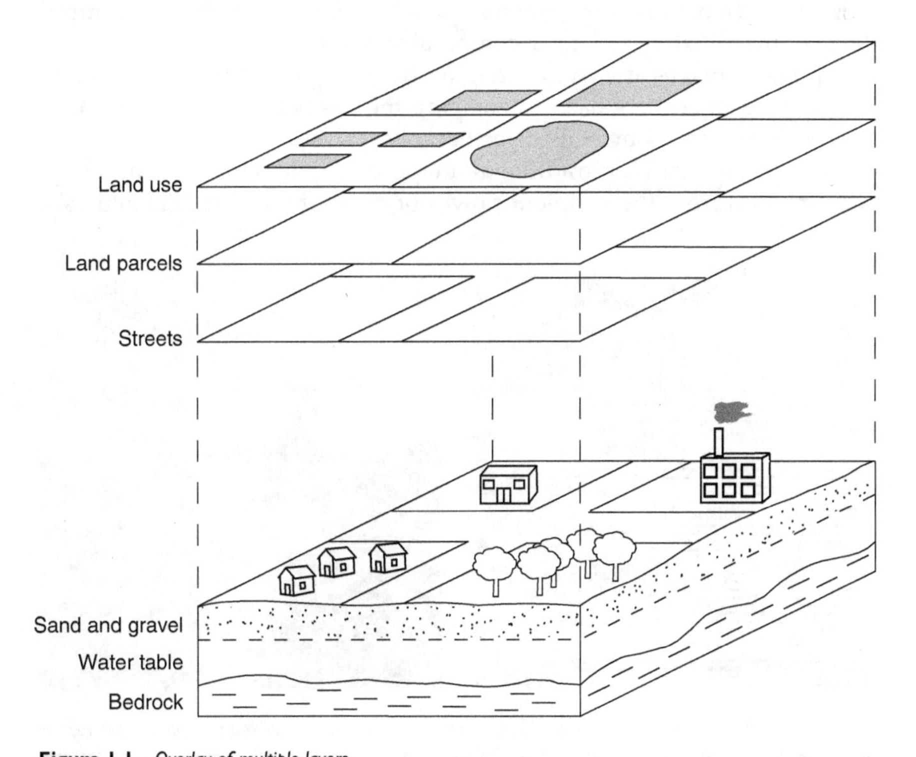

Figure 1.1 *Overlay of multiple layers.*
This process allows policy and decision makers to visualize possibilities and impediments associated with location of strategic facilities.

none of McHarg's initial analysis was done using a computer. Indeed, computers of the day were very primitive, and required massive physical and human resources to run. It is the metaphor of overlay, however, that was integrated into early GIS, and became the basis for a range of analytical techniques broadly known as "spatial analysis."

Spatial analysis is differentiated from "mapping" because it generates more information or knowledge than can be gleaned from maps or data alone. It is a synergestic means of extracting information from spatial data. Mapping, however, represents geographical data, with varying degrees of fidelity, in a visual form. It does not create more information than was originally provided, but does provide a valuable means for the brain to discern patterns, especially given that more than 50 percent of the brain's neurons are used for visual intelligence. In the early development stages of GIS, however, few people recognized the power of analysis, and the technology was generically referred to as "computerized cartography." As such, GIS made a very poor showing. Early computerized maps were very primitive compared to the exquisite product possible with manual cartography. Figure 1.2 illustrates an early computerized map juxtaposed to a comparable map of the same area. This comparison makes it easy to imagine the basis for initial resistance to GIS from geographers used to enjoying the aesthetic pleasures of maps manually produced by skilled cartographers.

The visual merit of traditional maps acted, however, as a decoy, a distraction from the incipient power of computerized spatial analysis.

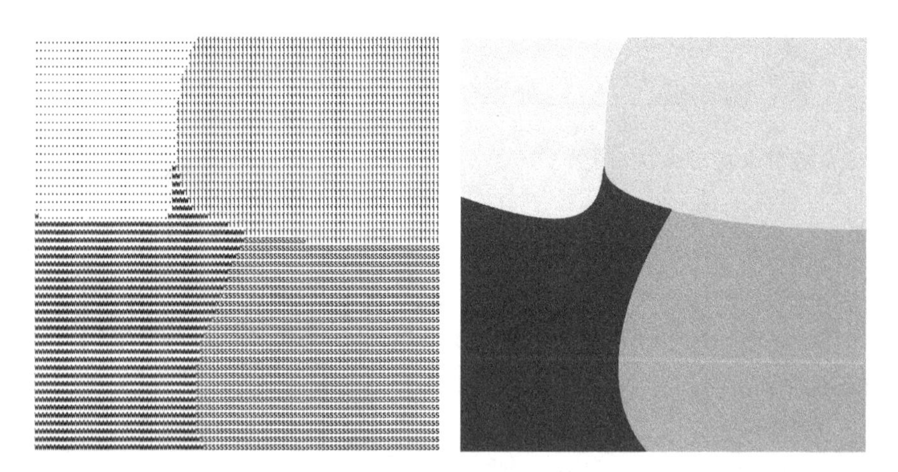

Figure 1.2 *An early computer display of differentiated spatial zones juxtaposed to a comparable map of the same areas using present-day technology. The graphical limitations of early computerized cartography inhibited the adoption of GIS methods for many geographers for whom the cartographic paradigm was paramount.*

That power was first explored in the late 1950s and early 1960s by researchers in the United States. Harold McCarty at the University of Iowa and William Garrison at the University of Washington were both experimenting with computational methods for analysis of large geographical data sets (N. Chrisman, 1988, personal interview). Influenced by the quantitative revolution and the development of computers, researchers began to develop tools that could be used to analyze and display spatial data.

One of the earliest computer cartography systems was developed in Canada, the brain child of Roger Tomlinson and Lee Pratt who met while sitting next to each other on an airplane (Tomlinson, 1988). Tomlinson had been using aerial photography to map forest cover in order to recommend locations for new growth. Lee Pratt worked for the Canadian Ministry of Agriculture. The Ministry wanted to compile land use maps for the entire country, maps that would describe multiple characteristics including agriculture, forestry, wildlife, recreation areas, and census divisions. Tomlinson suggested that they pioneer a computerized system in which land use zones were digitally encoded so that they could be overlaid with other relevant layers such as urban/rural areas, soil type, and geology. This happenstance meeting led, in 1964, to the Canada Geographical Information System (CGIS). The name of the system was bestowed by a member of Parliament – an instance in which sheer contingency cast a long shadow!

This Canadian version of the history of GIS is paralleled by efforts in the UK and United States during the same period. David Rhind (1988), a member of the UK Ordnance Survey has identified two streams of innovation in the development of GIS. The first was initiated by traditional cartographers who (slowly) began to recognize the merits of digitizing spatial information, and creating automated maps in a cost-effective manner. Parallel developments among quantitative geographers were initially quite separate. Brian Berry, Waldo Tobler, and Duane Marble in the US, and Tom Waugh and Ray Boyle in the UK began to develop algorithms and computer code to solve spatial problems. Their work became the basis of spatial analysis in GIS (N. Chrisman, 1988, personal interview.

In the US, the Harvard Graphics Laboratory was a tinderbox of the GIS revolution. Research at the lab established an efficient method for computerized overlay using polygon (vector) boundaries. The lab was populated by a host of researchers who continue to influence the development of GIS today including Nicholas Chrisman and Tom Poiker. A diaspora of researchers from the Harvard Laboratory in the 1970s contributed to the dissemination of GIS especially into the private sector. Scott Morehouse, a junior member left in 1981 to work for a company in

California called Environmental Research Systems Inc. (ESRI). At ESRI, Scott redeveloped the algorithm for vector overlay which became a cornerstone of the program ArcInfo®. This dispersion of ideas from the Harvard Lab was the beginning of one GIS identity: that linked to software packages, hardware systems, and technology in general (Chrisman, 1998).

The Messy Business of Digging For Roots: GIS' Intellectual Antecedents

The development of GIS, however, is not rooted solely in computer laboratories in the mid-twentieth century. It is arguably an outgrowth of attempts to automate calculation in the nineteenth century reflected in efforts, for example, to code population data for the US census in 1890. Pre-eminent GIS scholar, Michael Goodchild (1992), makes the point that GIS was developed during a period when information was increasingly being translated into digital terms and disseminated widely. If geographers hadn't explored the possibilities of digital manipulation of spatial data, other disciplines would have initiated the process. As it is, many roots of GIS are in disciplines other than geography including landscape architecture and surveying. Many GIS scholars regard GIS as an inevitable development, in the light of rapidly converging information technologies in a number of disciplines, combined with a recent history of spatially oriented, quantitative research questions in geography. An increase in scales of counting and analysis is part of a broader social and political movement toward enumeration and control of populations. Like all technologies, GIS is an outcome of both social and technological developments.

All disciplines have intellectual roots, or modes of thinking about phenomena that explain why certain methodologies are used, and certain knowledges privileged. Given that GIS is a relative newcomer to geography, one might think that it would be easy to nail down its intellectual antecedents. But the reverse is true. Although some human geographers claim that GIS is a direct descendant of the quantitative revolution, GIS researchers are loathe to accept this simplistic genealogy. They argue that its antecedents are more complex, comprising a number of threads which were, by circumstances of academic and technical progress, merged into GIS. Others regard GIS as a vehicle for quantitative models but profoundly more than a sum of techniques. Still other researchers argue that GIS transcended the quantitative revolution by incorporating visual intuition. There is a further sense that it is futile to categorize GIS' historical relationships, especially when they arguably began in the nineteenth century with the collection of statistical information about citizenry.

GIS was certainly one vehicle for the introduction of spatial techniques into the discipline but it piggybacked on quantitative methods. These two approaches were merged with the introduction of computer programming to solve spatial problems in the 1970s. Many geographers argue, however, that GIS is more than a vehicle to transport quantitative techniques into geography. According to Nancy Obermeyer, GIS bears the same relationship to the quantitative revolution that a calculator bears to mathematics. "On one hand, the operations that are clear-cut can be done more simply but...you still need to understand the models and the conceptual issues that underlie it" (cited in Schuurman, 1999a, 24). In other words, GIS draws upon models developed in the quantitative revolution but meaningful implementation still requires an understanding of how those models function in a spatial and algorithmic context. Having a GIS on your desk is not sufficient to implement quantitative models. Users are still required to understand how to frame the questions and wager the degree to which the question is appropriate in the context of the available data.

There is a divide between those who emphasize GIS' links to quantitative analysis, and those who regard it as an extension of mapping. Much early GIS simply involved using the brute force of computer cartography to map data distributions. David Rhind (1988) notes that there was a divide between people using the computer to *analyze* spatial data and those using it to *print* data in graphical form. Waldo Tobler, a legendary figure in both spatial analysis and cartography, used the computer to draw and calculate projections but remained a true cartographer in that he viewed transformation (spatial analysis) as a means to graphic representation, rather than an end in itself (Schuurman, 1999b). The argument is increasingly moot as, since the 1970s, output from analytical operations has been ported to printers for display – the basis of modern GIS.

Despite demonstrated antecedents of GIS in cartography and quantitative methods, there is an inchoate but emphatic sense, among researchers, that GIS extends quantitative techniques. By making them more accessible, many feel that it has imbued them with a more *intuitive* cast. One of the chief virtues of GIS is that it allows the visualization of spatial data as well as providing a means of utilizing fuzzy data. While the quantitative science prefers clear and precise "facts," GIS provides a way to include data that is not so pristine. It presents geographers with ways to visualize spatial arrangements and, in the process, recovers intuition from the wasteheap to which it was relegated during the quantitative revolution. Researchers in "scientific" visualization stress that it is the relation of graphical display to communication of information that distinguishes the methodology. The methodology is indeed superceded by the power of the image. This topic is treated with greater detail in Chapter four, and Figure 4.17 illustrates the influence that an image has in conveying information. In this example, the large spike in incidences of tuberculosis delivers information more powerfully than a table showing incidence of illness in various postal codes.

In GIS, visualization is emerging as a subspecialty that focuses on how humans interpret visual imagery, and algorithms for data manipulation and patterns of human-computer interaction. A surface map of elevations conveys a more easily interpreted feel for the landscape than a table that assigns an elevation to each grid cell for the same area. Visualization is used to manufacture meaning from data, through rendering it in image form. GIS incorporates ongoing research into geographic visualization but, more to the point, it is based on the very principles that have recently brought scientific visualization to the fore. Geographers have always used graphical representations to "see" spatial patterns.

GIS researchers perceive its visuality as a means of increasing the accessibility and meaning of spatial analysis. In a decision-making context, for instance, visual display often leads to *intuitive* conclusions about cofactors for a given incidence. This reliance on visual intuition constitutes a seemingly "unscientific" approach but it is one that finds increasing support in cognitive research which has demonstrated that people are able to discern information from visual display with greater facility than from tables or printed text. Furthermore, many scientists report that people "reason" using imagery. Visual images are processed by the viewer differently than numerical or textual output.

Despite recent incorporation of intuition and visualization into GIS' repertoire, it is difficult to dispute that there are "cultural affinities" between GIS researchers and quantitative geographers. Strenuous differentiation of GIS from "simple" quantitative analysis signals perhaps a reluctance to be tarnished with the same criticisms as have been leveled at mathematical modelers. It also points to a firm conviction, on the part of developers, that GIS surpasses the limitations of conventional analyses through its visuality. Consistent with a tendency to distinguish GIS from other strands of geography is a recent twist on its appellation. GIS now routinely refers to geographic information *science* rather than systems. The name shift points to qualities associated with the technology as well as its disciplinary context.

What Does the Acronym GIS Stand For? The Two Faces of GIS

Definitions of GIS tend to focus on the collection of hardware and software that are associated with the technology. A standard recital of what geographic information *systems* are might mention necessary components such as: methods of data input, analysis, mapping, and output associated with spatial data. Such definitions focus on a collection of practices, hardware, and software that have become known as GIS. Each of these algorithms, bits of metal, and computer code have their own ethnography, but they are so closely linked in the minds of their users as to form a "black box." The term black box was promoted in the popular literature (well, popular for academics) by Bruno Latour (1987) who argued that new scientific knowledge is at first disputed and references to it use copious citations to establish its legitimacy. As the concept – or technology – is better established, it is simply assumed to be true and good, and references and justification are no longer required. The term black box is suitable for one of GIS' identities – the systems identity. Most users, after all, who use a hydrological model embedded in ArcInfo® – a popular GIS program – don't question its legitimacy. Seldom does any one ask how their GIS software decided on the boundaries of the colored polygons that illustrate areas of different income level in a city. Nor is the spatial analysis routine that determines daily delivery routes for a courier company likely to be disputed. GISystems are assumed by the vast majority of users to produce true results.

Close by in a parallel universe, geographic information *science* is concerned with precisely these questions. GIScience is, in the simplest sense, the theory that underlies GISystems. It took several decades, however, for this alternate GIS identity to emerge. By the beginning of the 1990s, a sense prevailed among many academic researchers that GIS had forged new intellectual territory. This intimation was first given substance in a keynote speech given by Michael Goodchild, Professor of Geography at the University of California at Santa Barbara, during the July, 1990, Spatial Data Handling conference in Zurich and again at the EGIS (European GIS) meeting in Brussels in April, 1991 (Goodchild, 1992). In each of these addresses, Goodchild noted that the GIS community is driven by intellectual curiosity about the nature of GIS. He argued that it behooves researchers in GIS to focus on fundamental precepts that underlie the technology rather than the application of existing technology. Furthermore, he argued that there are unique characteristics of geographical data, and problems associated with its analysis, that differentiate GIS from other information systems. These properties include: the need to develop conceptual models of space; the sphericity of spatial data; problems with spatial data capture; spatial data uncertainty and error propagation; as well as algorithms and spatial data display. Given the distinctiveness of geographical data analysis and a growing community of researchers dedicated to solving technical and theoretical problems associated with GIS, Goodchild argued that "GIS as a field contain[s] a legitimate set of scientific questions" (cited in Schuurman, 1999b). Questions about the underlying assumptions written into the code that comprises GISystems are the basis of GIScience.

A GIScientist would indeed question the premises of a hydrological model. She might ask who devised the model? How well does it work in a glacial environment as opposed to a wetlands? Or, is the model designed for use with vector (polygon) or raster (gridded) data? Other GIScientists are interested in how boundaries are defined. How do different input parameters or measurement systems lead to different boundary definitions, and how do these vagaries affect the results of GIS analysis? Network analysis routines that optimize delivery or repair routes are also subject to deeper investigation. A GIScientist is likely to try and ascertain whether certain neighborhoods are better served than others, and whether travel times accurately reflect changing weather and traffic conditions. These types of questions strike at the efficiency and legitimacy of current GISystems algorithms, and their resolution will greatly increase the reliability of GIS for the average user. They don't represent, however, the entirety of GIScience.

Every stage of GISystems from spatial data collection and input, to storage, analysis, and, finally, output of maps is based on the translation of spatial phenomena into digital terms. At each step of GIS, data are manipulated for use in a digital environment, and these, often subtle, changes have profound effects on the results of analysis. Each of these transformations involves a subtle shift in the representation of spatial entities, and accounting for these modifications and their implications is an important part of GIScience. Physical and social information about the world, once in digital form, is often manipulated and analyzed *in order to* correspond to the researcher's interpretation of the world. Thus, it is of fundamental importance that GIScientists understand how to monitor and account for the effects that transformations have on data. Finally, GIS researchers must understand how to present analyses such that their visual display is consistent with database results.

The work of GIScientists begins even before data are digitally encoded. Spatial phenomena must be delineated and classified in preparation for input to data tables. Classification systems, however, must be compatible with data tables, and this acts as a constraint to the development of categories. Many spatial phenomena manifest multiple characteristics, but not all of them can be included in a database or the data would be infinite. The manipulation of data depends on the attributes that are recorded, or the objects that are defined. Different community boundaries, for instance, will render different results in an assessment of population health. Visualizing GIS results is likewise vulnerable to the vagaries of the digital environment, and must be consistent with human capacity for perception. At a small scale, for instance, only a limited number of attributes can be displayed or the map becomes overcrowded. At a larger scale, a greater number of attributes can be accommodated. Each of these issues has a bearing on how spatial data are analyzed and interpreted.

In the broadest sense, GIScience is the theoretical basis for GISystems, and its research purview is the representation of spatial data and their relationships – in terms of bits and bytes. Working in a digital environment is akin to speaking another language that uses fundamentally different building blocks. If we think of the English language as being composed of 26 letters that can be combined in various ways to form words, sentences, and ideas, then GIS is based on two letters (well, digits – zeros and ones) that can be combined and manipulated to represent and analyze geographical phenomena and relationships. But the environment and rules associated with manipulating geographical objects are quite different from those we are accustomed to using for text and conventional graphics. GIScientists explore how spatial objects become digital entities, what effect that transformation has on their ontologies, how to represent different epistemologies within GIS, how to model relationships between spatial entities, and how to visualize them so that human beings can interpret the results. This pursuit draws on and extends developments in data modeling, computer science, cognition, scientific visualization, and a myriad fields that have emerged in response to information systems.

GIScience is not limited, however, to process-oriented issues. It is engaged with how people represent their geographical environment, and who has the authority to represent space. Public Participation GIS (PPGIS) studies and engages with nonprofit groups and nongovernmental organizations who use GIS to represent themselves, and advocate for change. Other GIScientists address questions about feminism and GIS, and whether the technology is inherently gendered. Stacey Warren (2003) explains that PPGIS and feminism and GIS allow us to move the focus from analysis and representation in GIS to one that views the technology as a "collaborative process that involves both people and machinery." This emphasis on social interactions between users, affected populations, and technology is evident in the growing number of *Critical GIS* scholars who have merged emancipatory agendas and theory from human geography with GIScience.

Developers and researchers postulate that GIScience transcends mere information *systems* and allows users to ask questions about spatial relations that were previously impossible to pose. Its champions argue that geographic information *science* extends spatial analysis by virtue of enhanced processing power that allows data-intensive analyses to extend their geographical breadth. They claim that GIScience is a means of investigating previously obscured spatial relationships and contingencies. There is a tension between GIS scholars who view the technology as an emergent phenomenon, capable of initiating a shift in scientific methodology and other geographers who view it simply as a vehicle for concepts that emerge from geography. It is, of course, both. Moreover, GISystems are the medium for ideas that emerge from GIScience. This text uses the acronym GIS in most instances for simplicity to refer to both systems and science. This conflation of terms reflects both the interrelatedness of two pursuits, as well as the fuzzy boundary between them. In cases where their differentiation is important, the distinction is made.

Data In, Information Out: Common Ground Between GIScience and GISystems

Despite having elaborated on the distinctions between GIScience and systems, the same practices define them. GISystems incorporate processes such as classification, digital encoding, spatial analysis, and output into software, while GIScience provides the theoretical bases and justification for the *way* that these processes are executed. Both start with, and are dependent on spatial data. After the initial problem of identifying which spatial entities (such as houses, communities, forests, roads, or bridges) need to be defined as data, the information must be collected and classified. Classification is a messy business with different categories leading to alternate representations of the same spatial objects (see Chapter 3). People disagree about the definition of the boundaries of spatial objects, and even more strongly about how to put them in categories. You might ask the question, for example, where does the mountain end and the foothills begin? If this can be established, then you are still left with the problem of how to divide the mountains into categories. Is the 1,000 m elevation mark a critical divider, or should all mountains under 5,000 m belong to the same category? These discussions become quite heated when resources are involved. If communities with an income level below a certain mark are eligible for federal funding for health clinics, then the way that income is defined becomes a matter of some importance.

The territory that boundaries encompass has equal bearing. Boundaries drawn around communities yield very different results at different levels of aggregation. Use of Enumeration Areas or EAs (Enumeration Districts in the UK) as the basis for analysis of income levels will yield very different results than using Central Metropolitan Areas comprised of multiple EAs. GIS software is also best suited to crisp, linear boundaries, which creates a predicament for researchers who are not quite certain how to draw the line between, for example, black bear and grizzly bear habitats. Indeed, a fundamental challenge for GIScientists is to find ways to represent the fuzzy boundaries that characterize geographical areas and events using the crisp lines favored in GIS.

Further challenges are associated with modeling spatial phenomena using GIS. Spatial analysis and modeling are increasingly used to predict outcomes, and plan for future development or natural hazards. In the past, GIS was used primarily to manage data, and map distributions. This capacity has been extended by the ability to model interactions among different attributes (characteristics) of the spatial objects, and use this information to predict future events. Land use managers and city planners, for instance, use GIS to study future urban growth based on multiple factors such as density, socioeconomic indicators, geographical constraints (e.g., is the city bounded by mountains or ocean?), road networks, and present land use. Once data are classified, spatial boundaries determined, and analysis complete, the results must be visualized so that users can interpret the information.

Geographic visualization refers both to traditional cartography and to the ability to express knowledge about space and spatial relations in a visual form. The power of GIS emerges partly from its capacity to make visual spatial relationships, and to picture spatial objects in a way that allows users to interpret pattern. Rather than generate tables listing the census tracts associated with children at high risk of contracting Hepatitis A in their preschool, a GIS graphically displays the census tracts, color-coded based on level of risk. The value of visual display in assessing pattern associated with the spread of disease in illustrated in Figure 1.3. At the analysis level, there is no perceptible difference between the statistical results and GIS. The visuality of results, however, allows for intuitive *or* structured exploration of cofactors. The most famous example of visual intuition related to mapping is that of epidemiologist's John Snow's hypothesis that Cholera incidence, during the 1854 outbreak in London, was highest in the vicinity of public wells. Figure 1.4 illustrates the distribution of deaths from Cholera and public water pumps in the Soho area of London in 1854. Based on this map, Dr. Snow reputedly discerned that the use of public wells was linked to Cholera. This conclusion was not straight forward as there were several buildings with high population density in the vicinity of the Broad Street pump in which there were no deaths. Snow relied on his local knowledge to visit the Poland Street workhouse, for instance, and ask from which pump the inhabitants drew water. It turned out that the workhouse had its own well, and none of the 135 inmates had visited the Broad Street pump.

This story demonstrates the value of local knowledge used in conjunction with maps to discern patterns. Visualization in conjunction with GIS

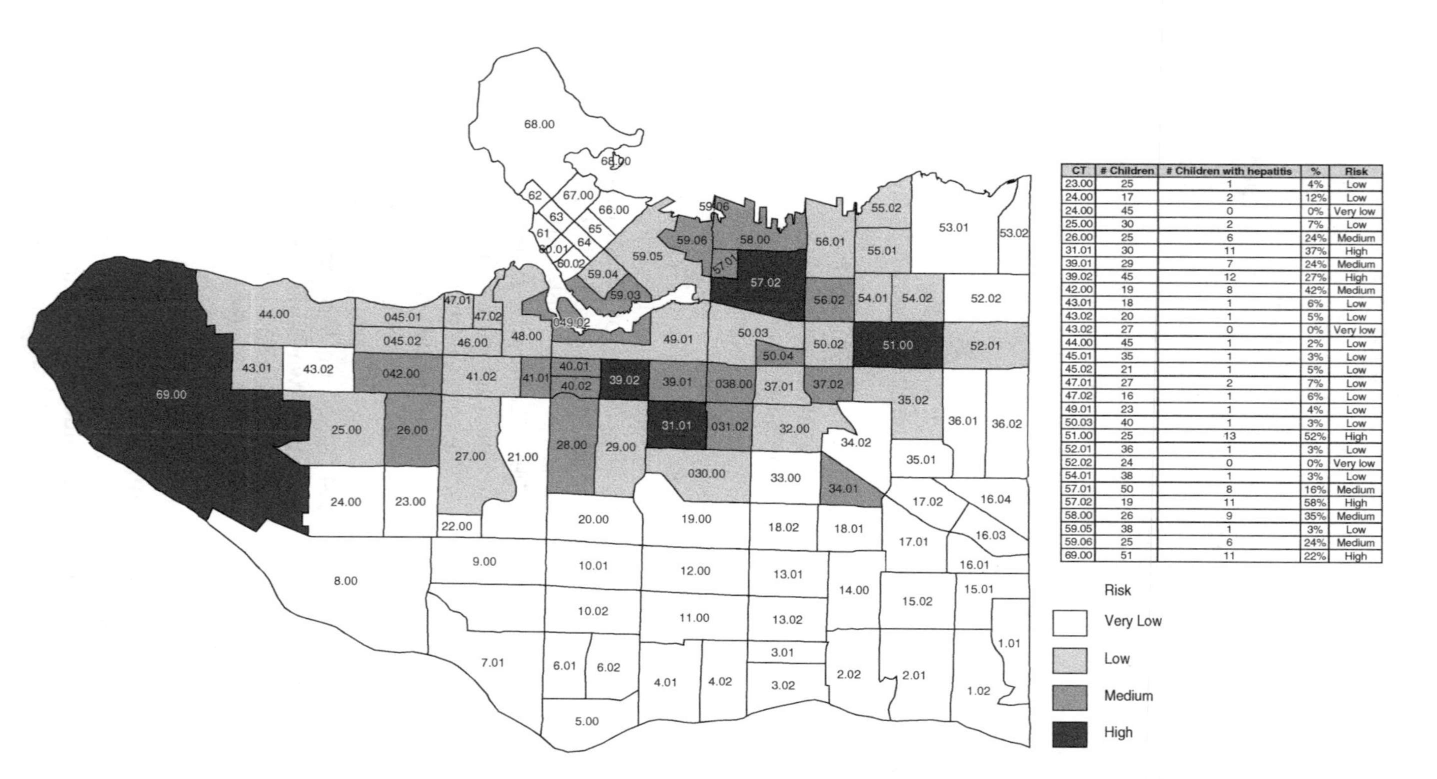

CT	# Children	# Children with hepatitis	%	Risk
23.00	25	1	4%	Low
24.00	17	2	12%	Low
24.00	45	0	0%	Very low
25.00	30	2	7%	Low
26.00	25	6	24%	Medium
31.01	30	11	37%	High
39.01	29	7	24%	Medium
39.02	45	12	27%	High
42.00	19	8	42%	Medium
43.01	18	1	6%	Low
43.02	20	1	5%	Low
43.02	27	0	0%	Very low
44.00	45	1	2%	Low
45.01	35	1	3%	Low
45.02	21	1	5%	Low
47.01	27	2	7%	Low
47.02	16	1	6%	Low
49.01	23	1	4%	Low
50.03	40	1	3%	Low
51.00	25	13	52%	High
52.01	36	1	3%	Low
52.02	24	0	0%	Very low
54.01	38	1	3%	Low
57.01	50	8	16%	Medium
57.02	19	11	58%	High
58.00	26	9	35%	Medium
59.05	38	1	3%	Low
59.06	25	6	24%	Medium
69.00	51	11	22%	High

Figure 1.3 *Graphic display of one scenario for incidence of Hepatitis A in daycares in the Vancouver area. Note that areas of high incidence are much easier to discern from the map than the table.*

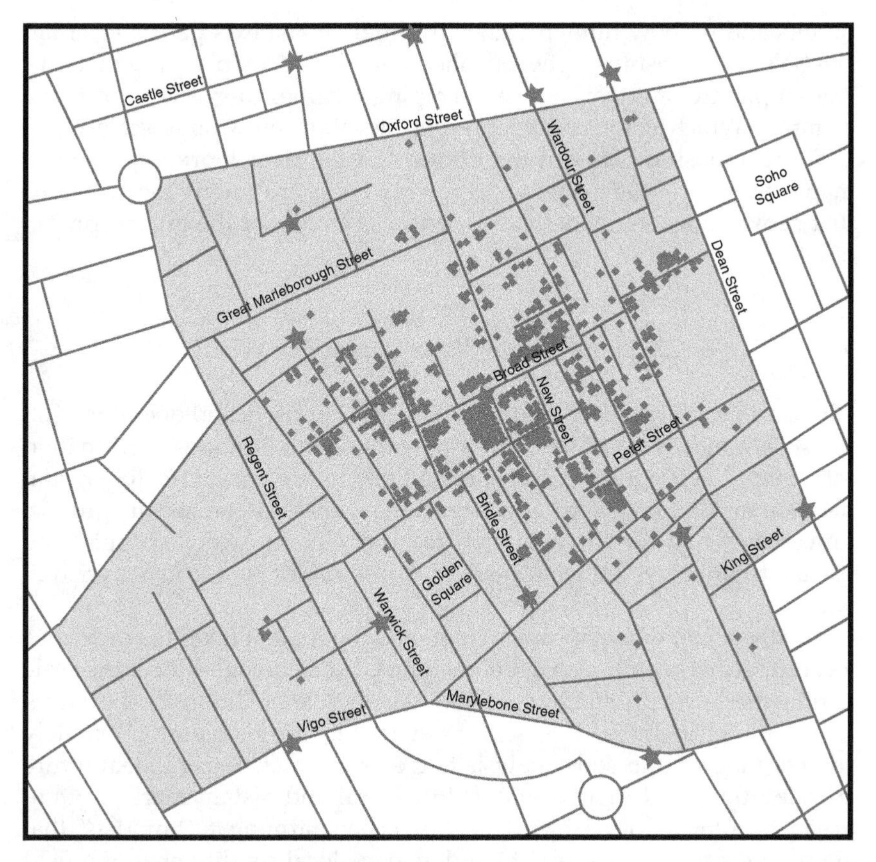

Figure 1.4 *Cholera outbreaks in London in 1854.*
The stars represent pumps and the dots Cholera cases. By making a map of each incidence, Dr. John Snow was able to make the famous connection between water pumps and the spread of Cholera.

is linked to a trend in science toward using visual displays to understand pattern, and ultimately cause and effect. An example of the power of visuality in science is the discovery/production of the double helix structure of DNA, based on images developed through X-ray crystallography. More recently, the geography of the human genome has been mapped to assist researchers in understanding relationships between chromosomes.

Links between visual intuition, knowledge discovery and computer technology have been the subject of intense research during the past decade. Generating a reliable visual display is, however, much more complicated than it may appear. At the most primitive level, each spatial object must be translated into rows of pixels with varying degrees of color, hue, and saturation. But visualizing spatial data also entails understanding how human beings in different contexts perceive certain symbols, relationships between phenomena, and map representations. Does a picture of a teepee mean camping facilities to everyone in every country? Which colors best represent elevation on a large-scale map? Is the relationship between the bridge and the river more important to map readers than their precise geometry? These are among the questions that geovisualization experts must address as part of the greater project of GIS.

GIS in the World: Who Uses It For What?

GIS has a pervasive reach into everyday life. For users and operators, GIS provides a means to convert data from tables with locational information into maps. Subsequent GIS-generated maps are the basis for spatial decision making in government agencies, businesses, community groups, universities, and hospitals. But the reach of GIS far exceeds people who use the technology. It affects the lives of millions of people in a myriad of ways.

What you eat, where it comes from, and the route it takes to reach your local supermarket are each dependent on GIS technologies. As large-scale agribusiness has proliferated, so too has the role of GIS in food production and agriculture. Business farmers regularly combine remotely sensed imagery and soils analysis to create visualizations of ideal future crop locations and their relationship to local and distant markets. Quotidian farming is often based on "precision farming techniques" that allow the farmer to respond to and analyze local conditions in the field with pinpoint accuracy. For instance, a section of a wheat field might have blight. The area circumscribed by the blight is inventoried using global positioning systems (GPS), and then combined with other layers such as soil type, soil chemistry, wheat variety, pesticide load, and irrigation information to determine why that particular section is under duress. Likewise, data about grazing are used to assess the number of beef cattle the land will support based on a given area of pasture. Crop management includes planning to protect vulnerable crops from frost, fires, and over-precipitation or drought. GIS is used to model each of these factors and provide risk factors associated with each depending on the crop and type of farm (e.g., organic or conventional; hand-harvested or machine-picked). Once harvested, crops need to reach a wide range of markets depending on purchase pricing, local preferences, and the cost of transport. Finally, modern farming is sensitive to markets. GIS is used to profile markets, pricing and related transportation costs in order to develop an optimum model for matching crops to consumers.

Municipal management, like farming, has become a high-tech field that is dependent on GIS for delivery of services. A brief inventory of spatial data held by almost every municipality includes property outlines with survey points, tax assessment values, township and country boundaries, roads, waterways, public transportation routes, bicycle paths, aerial photography, park lands, public buildings, and waste collection routes. Each of these spatial entities is associated with a particular GIS functionality. For instance, tax assessment values are linked to individual houses, and are used to evaluate levels of service associated with particular neighborhoods – as well as to keep track of the payment of taxes. Road files, including surface material, embankment, and grade are combined with elevation, weather, and traffic volume and load data to determine which roads are likely subject to accelerated degradation. When roads require repair, closures and reroutings are designed to minimize traffic disruption – though this is seldom clear when you are sitting in stalled traffic. Encouraging bicycle use and green commutes is the goal of an increasing number of urban municipalities. Since 1993, the city of Vancouver, British Columbia, has designated a 135 km bicycle network throughout the city. Since Vancouver has only 5 km of dedicated bicycle path, GIS is used to estimate traffic volumes of both modes of transport during peak commuting periods in order to determine relatively safe bicycle venues. Accessibility of different neighborhoods to parks or public services such as libraries is determined through GIS queries. Waste collection routes are designed using GIS network analysis to reduce exposure of pick-up trucks to traffic, and to optimize the amount of waste gathered on each collection route. This description is an attenuated account of the degree to which GIS has become instrumental in planning our cities.

Urban life is also reliant on GIS in more subtle ways. Pervasive and complex networks provide power, fuel, and water to town and city dwellers. The electrical grids that deliver power are designed and managed using GIS. Each circuit is mapped, and its direction recorded. Circuit can be traced down to the individual customer, and load concentrations can be visualized on a house-by-house basis or for the entire neighborhood. When a circuit needs to be closed down, these data are used to examine all feeding directions in order to switch locations and minimize electrical outages. Specialized software is used with these data to balance transformer loads and minimize loss of power as it seeps through the lines. Recent trends toward privatization of public utilities in Europe and North America have increased pressure to achieve greater efficiencies. GIS has played a role in this trend by offering fully functional systems that not only manage infrastructure, but create virtual models for switching and control systems. These allow managers to test complex scenarios for delivery and load including incorporation of "cogenerators" or small businesses that sell spare electrical capacity back to the main grid. The water reservoir and distribution system, natural gas fuel lines, and telephone and cable lines are similarly GIS based and managed.

G-commerce or e-commerce facilitated by GIS has burgeoned as web-based sales proliferate. G-commerce is based on mapping and data analysis tools that allow businesses to construct business-to-business (B2B) and business-to-customer (B2C) portals. A typical B2C portal is illustrated by Amazon.com which sells books, music, and even pharmaceutical drugs directly to consumers in their homes. B2B portals are just as common; they are the basis for "just-in-time" delivery systems in which production is wed ever more closely to sales in order to avoid long shelf lives for products – and delayed revenue. G-commerce also provides marketers with the tools to analyze data on customers, sales, and performances using socioeconomic and "lifestyle" data. These data are used to visualize consumer trends, and detect opportunities for increased sales. This trend contributes to what Mark Poster (1996) has called the creation of 'digital personae' in which each individual is incompletely described in government and marketing databases based on frequently scanned digital data and derived consumer profiles. These data and accompanying profiles are necessarily incomplete and result in only a rough approximation of each of us. They are the basis, however, for much marketing and determine where new retail outlets are opened, and whether you receive a given flyer in your letter box.

The use of digital data on individuals and communities is not used only by private firms; rather it constitutes the basis for e-governance or electronic governance. E-governance is proliferating as federal and provincial governments begin to use the web to deliver services and allow public access to information. E-governance has an a-spatial, administrative ring to it, but it is powered by GIS and related "spatially aware" software. At the municipal level, e-governance entails access to survey lines, property definitions, and tax assessment information. Public notices are web-posted, and forms for everything from dog-licensing to tendering of construction contracts are managed on-line. At the state or provincial level, e-governance is poised to become the vehicle for automobile registration and other services including campground reservation, passport renewal, postal services, and plebiscites. The appeal of e-governance is the promise of more efficient and transparent delivery of services. Its success is dependent, however, on high-levels of web-access which is still not a reality in most countries. Interestingly, India is at the forefront of e-governance technologies and implementation. This speaks to the remarkable intellectual capital the country has as well as the ability of technologies to "leap-frog." The proliferation of cell-phone use in sub-

Saharan Africa by people who never owned a land-line is one example of technology leap-frogging. In the case of India, proponents of e-governance argue that it is a means to eliminate high levels of corruption in the civil service while optimizing the delivery of government services. Detractors counter, however, that e-governance is a means of centralizing power in the hands of a few, and that it lends itself to the indiscriminate collection of digital data about individuals in the absence of privacy restraints. These arguments aside, e-governance is being actively pursued by almost every level of government in many countries. The technology to do so is dependent on spatial data and GIS functionality.

Clearly GIS is interwoven with the fabric of every day life. Understanding the computational and intellectual basis for this technology is an excellent first step toward a better comprehension of the technological bases for modernity. This understanding is a starting point for insights into how the digital realm has come to organize and control so many functions of modern society. The rest of this book sets out to accomplish this task by examining not only GIS the technology but its intellectual and disciplinary ties.

In Chapter 2, the relationship between GIS and human geographers within the discipline of geography is explored as a way of delineating their shared intellectual territory. Explanations for past stormy relations are offered from the perspectives of both disciplinary niches. Many of the initial differences between GIS scholars and social scientists are linked to epistemology or the formal and informal perspectives that inform research methodologies. While epistemology of implementation and development affects GIS, there are myriad contextual factors that influence the technology. The second part of Chapter 2 examines ways in which the development of GISystems and GIScience have been shaped by intellectual traditions, language, and political pursuits.

Using GIS requires data – or information – as well as appropriate software. In fact, data are the primary determinant of relevance for GIS analysis. Students and users of GIS are often captivated by the power of the software, and presume that data are appropriate by virtue of their existence. Chapter 3 is concerned with spatial data including the politics of collection and their relationship to representation, how data are organized, and the challenges of sharing data. The discussion of data ranges from the sociopolitical contexts of collection to the technical challenges of interoperability between data sets. The chapter concludes with an example of data collection and sharing that demonstrates the constraints of the technology, and the politics of implementation.

Data are the servant of analysis in GIS rather than ends in themselves. Chapter 4 delves into the operations that give GIS its power: the constituents of spatial analysis. The early part of the chapter is necessarily devoted to explaining the basis for common spatial analysis operations, their parameters, and the logic upon which they are based. The latter section is devoted to working examples of GIS in environmental management and population health. Finally, the rationalities of GIS analysis are examined with an eye to reinforcing the notion that GIS like statistics strengthens particular actors and agendas.

In the final chapter, the distinction between GISystems and GIScience is revisited in order to afford the reader a more nuanced notion of what everyday work in each of these niches might entail. The potential of GIScience research to enhance the scope of representation afforded by current GISystems is described by providing a brief description of two small, but significant areas of current research: ontologies and feminism and GIS. Both of these areas are of interest to human geographers because they share common literatures and ideals. In the case of ontologies research, the goal is to enable GIS – as a form of representation – to better model the world based on multiple perspectives. Feminism and GIS incorporates and furthers the goals of feminist politics by incorporating and changing GIS to better serve as an ally in these endeavors. The concluding section reiterates the interrelatedness of GISystems and GIScience, and the value of both the discipline and pursuit of geographical knowledge and representation.

2

GIS, Human Geography, and the Intellectual Territory Between Them

Both the systems and science component of GIS constitute part of an intellectual territory in which certain assumptions are privileged over others. Intellectual pursuits and their technological products are not isolated in time or practice. They are, in turn, influenced by the disciplinary cultures and ethos. In the first part of this chapter, the relationship between human geographers and GIS scholars is explored. This relationship is relevant to the development of GIS technology and applications as well as to the discipline of geography; it speaks to differences in intellectual culture and practices between many human geographers and GIS scholars. Moreover, it points to ways in which these two disciplinary niches have influenced each other, and continue to affect each other. The second part of the chapter introduces some aspects of the unique intellectual territory of GIS including the bases for representation of space and spatial entities as well as the philosophical space in which these representations emerge. Descriptions of the primary data models used in GIS illustrate the mediating role of GISystems in representing the spatial world. Technologies are never outside the social, and the final section of the chapter illustrates how social influences on GIS can be detected in the technology, and how such insight can be used to imagine a better GIS.

Geography began as a loose collection of scientists and empiricists interested in the physical nature of the earth's surface as well as the role of geography in constructing politics and shaping behavior. It didn't constitute a discipline proper until at least the beginning of the twentieth century. Human geography as an academic discipline began to coalesce in the following decades led most famously, perhaps, by American Carl Sauer who used multiple intellectual tools to understand cultural dimensions of spatial change. By World War Two, human geographers were a significant constituent of geography. By contrast, GIS wasn't named as such until 1964, though its roots are buried in the quantitative revolution and a long history of cartography. These two subdisciplines of Geography have little common ground – on the surface – and this has led to a rocky relationship in the recent past.

Mind the Gap: The Distance Between Human Geography and GIS

GIS and human geography occupied separate spheres of geography without much public interaction until the late 1980s when some human and cultural geographers turned their attention to GIS. Early critiques of GIS by human geographers focused on perceived methodological and epistemological shortcomings of the technology. GIS researchers were often defensive, and considerable jousting between the two subdisciplines ensued (Schuurman, 2000). Given that different people picked up the debate at various times, and others heard about it second or third hand, it is worth reviewing them precisely because debates about GIS have had effects on the discipline of geography *as well as* the way that GIScience structures research questions. These exchanges also point to the historical distance between human geography and GIS, a distance, that is increasingly bridged by cooperative efforts between researchers in both groups.

Published evidence of tension between these two camps was first evident in 1988 when the President of the American Association of Geographers (AAG), Terry Jordan (1998), characterized GIS as a "mere technique." This sentiment was common at the time, and reflected the prevalent sense that GIS was simply an automated form of cartography, bereft of intellectual substance. Subsequent dissent between human geographers and GIS researchers took place in the pages of *Political Geography Quarterly*. There was sentiment on the part of human geographers that GIS, while well equipped to manage information, is inadequate in the realm of knowledge production, concerned with facts but incapable of meaningful analyses. GIS was presumed devoid of theory or abstractions, and characterized as a geography based on "facts" rather than "real" knowledge. Moreover, GIS scholars were alleged to be steeped in positivism and naïve empiricism (see below). Such a geography, it was intimated will surely not last long.

Michael Goodchild, a prominent GIS researcher, responded to some of the initial complaints. Rather than extol the virtues of GIS, Goodchild argued that while the technological structure of GIS is controlled by computer science, the development of GIS in geography has led to the realization that databases and spatial analysis are subject to uncertainty, and can be inaccurate. He suggested that GIS has made its own limitations an integral part of its research for decades. Furthermore, he argued that GIS is most useful precisely when it is guided by geographers, and that GIS is designed to be used in conjunction with knowledge rather than a substitute for it (Goodchild, 1991). Goodchild defended early aspersions on the intellectual integrity of GIS, but the debates continued to proliferate within the pages of *Environment and Planning A* which published a number of articles exposing hostility between members of the GIS community and human geographers.

Heated opinions and barbed comments marked Stan Openshaw's contribution to debates over the value of GIS. Openshaw intimated that people in GIS have felt censured by other geographers and "their misinformed speculation about what GIS is and does, and how it either fits or does not fit comfortably within geography" (Openshaw, 1991, 621). He berated critics of the technology for qualitative methodologies that had little relationship to maps or spatial processes. Openshaw was understandably defensive of GIS, and felt that human geographers with little or no understanding of digital processes were in a poor position to comment on the merits of GIS. Openshaw's arguments were inflammatory, but they accurately reflected tensions within the discipline, and were refreshingly honest about the stakes involved.

While initial critiques of GIS covered a range of perceived shortcomings, positivism or, more generally, epistemology emerged as a basis for the scrutinizing of GIS. Human geography critics felt that GIS failed to accommodate less rational, more intuitive analyses of geographical issues, and that its methodology, by definition, excluded a range of inquiry. GIS scholars meanwhile saw the value of their techniques being denigrated without really realizing why. There was a sense of puzzlement that human geographers refused to acknowledge the predictive or explanatory power of GIS as a way of offsetting critiques. An either/or theme to discussions about methodology prevailed. The debate continued with vigor for three years until several GIS and human geographers arranged a conference to bring together antagonists and defenders of GIS at Friday Harbor in Washington state in November 1993. It marked the beginning of increased cooperation between the two groups.

Friday Harbor also initiated a shift in the tone of the debate. The group assembled at Friday Harbor brought to the table a range of perspectives about GIS thus creating an opportunity for intradisciplinary communication which had thus far eluded debates over GIS. Conference attendees from human geography were able to articulate their concerns about the development of GIS. The first was that technological design and logic have far-reaching and lasting effects. They privilege certain conceptualizations of the world. A second concern was that GIS development is presided over by private sector firms, and that the software is designed to solve corporate problems (e.g., how to route courier deliveries) rather than address social inequities. A third concern was that GIS is inaccessible to most people in the world, and, even if everyone in the world had access to it, it would continue to represent only a very limited, linear type of problem solving.

One of best known collection of critiques of GIS was *Ground Truth*, edited by John Pickles, a participant at Friday Harbor, and one of the most prominent critics of GIS (Pickles, 1995). *Ground Truth* was a project originally developed in conjunction with the late Brian Harley. It was theoretically influenced by the ground-breaking work of Harley on the relationships between maps and power. Harley claimed that maps have always been a mechanism for depicting and producing social relations. It is taken for granted, he maintained, that the king's castle should be depicted in a large size on a feudal map while whole clusters of feudal cottages are absent. The map is not a neutral representation of territory, but a representation of social relations. Pickles extended this analysis to GIS, declaring that GIS software and research programs are marketed with the promise of being able to enhance understanding and increase peoples' control over their own and others' lives a process that he referred to as the "colonization" of every day life. GIS practitioners were not, however, ignorant of the ways that maps could be used in the interests of power. Mark Monmonier's book *How to Lie With Maps* had tackled precisely the same issues, illustrating that maps are a means to exercise and enforce relations of power (Monmonier, 1996).

Critiques of GIS had, by 1995, taken Harley's analysis a step further, asserting that GIS not only represents but perpetrates certain relations of power. This view corresponded to sociologist John Law's (1994) contention that the vision of modernity has been accompanied by technologies used to create and enforce order within societies. The purported propensity of GIS to order society has historical roots. Law noted that between 1400 and 1800, Europe witnessed the introduction of new approaches to social organization in which maps were crucial for the representation and imposition of the "truth." Power may be internal to maps, but critics linked it explicitly to GIS. Harley's legacy was an abiding interest in the relationship between maps and power among cultural geographers. This was consolidated with the rise of GIS and a perceived need to gain some control over its effects within the discipline of geography.

Friday Harbor, while bringing together some critics and scholars of GIS, ironically marked a split within GIS between practitioners who responded to the language and content of the critics, and those who thought that it had very little to do with their own work. But, if GIS practitioners could not be homogenized, then neither could their critics. By the middle of the decade, differences in politics and strategy between critics were emerging. A number of human geographers and

GIS researchers had started to focus on the means of practical intervention. Public Participation in GIS (PPGIS) had become one venue for democratizing the technology, and a number of critics and researchers were engaged with community groups and nonprofit agencies who wanted to use GIS to "counter-map," or produce alternative representations of their communities and local issues in order to challenge corporate or state agendas.

At the same time, there was increasing recognition that technology was not good or bad, but part of a social process that evolves over time. Michael Curry (1997), a student of geographical thought, argued that GIS is not a *bad* technology that will expose our personal habits through large scale manipulation of spatial data. Rather historical and contextual meanings of privacy have shifted in response to technologies. His set of observations on the interdigitization of law, technology, and culture disrupted a pattern of laying the onus of responsibility on technology. Instead he sought to unravel the complexity and contradictions inherent in digital representation and, more to the point, explored ways in which digital individuals can reinvent themselves to accommodate the sure knowledge that they are incompletely inscribed in a database.

By showing that what is considered public or private is negotiated in the discourse of jurisprudence and politics, Curry shifted the onus of responsibility from GIS to its social context, explicitly invoking possibilities for resistance in every system, digital or analogue. He argued that we must see ourselves as responsible for our identities whether they be real or virtual. By acknowledging the culpability and, conversely, the power of individuals to adjust to social conditions, he shifted the onus from GIS to a complex matrix of juridical, cultural, political, and scientific realms from which it is produced and in which it operates. The important contributions of critics of GIS was to illustrate possible oppressions supported by GIS technology. The counter to this insight is that technological tyranny can be resisted at an individual and social level because power is flexible and circuitous.

In the present, both critics and defenders of GIS are better informed about the agendas and implications of each others work. A willingness to integrate dialogue and debate over the effects of GIS as well as its epistemological bases is also well-established and supported by institutional structures. But this recent history of debate over the value and integrity of GIS marks important territory. It speaks to the discursive divide between these two fields. Language is never absolutely clear and this is especially true when talking across a gulf like the one which separates GIS from its critics. Words such as *epistemology* or *ethics* may be interpreted differently by GISers than their critics. While social theorists may complain that GIS researchers don't use words like positivism, epistemology, or ontology in the way that they are intended, the converse is also true. Human geographers often use terms like *mapping* and *space* when discussing cultural or social phenomena. Ideas such as "cultural space" and "knowledge maps" employ a complex and nonlinear transformation to the physical space and the two are ultimately "unmappable." Scientists have attempted to mitigate this problem by insisting on explicit definitions, on *formalization* and information technologies to tie users to precise categories. Difficulties integrating different scientific classification systems speak, however, to the futility of relying on language for precision. Neither science nor social science can claim perfect communication. Mediating between the discourses of GIS and human geography is an attempt to dispel some of the misunderstandings that have separated GIS from its critics. Philosophical issues of epistemology and ontology dominated discussions between human geographers and GIS scholars, and their elucidation with respect to GIS is a good place to begin.

Epistemology and Ontology in GIS

Geographical relations are considered and depicted by both human geographers and GIS researchers along two philosophical axes: *epistemology* and *ontology*. Understanding how GIS treats these philosophical issues is one way to understand and mediate between the intellectual territories of GIS and human geography. But first, a brief explanation of the terms, and how they are interpreted in GIS. Epistemology refers, in the broadest sense, to the methods that we use to study the world, and the lenses that they entail. It refers to the perspective that a researcher uses to interpret entities and phenomena. Ontology refers to what something really is, its foundational essence. A geographer studying a forest fire uses an epistemological lens to interpret the phenomena (the fire) and the spatial entities it affects (the trees). The ontologies of the forest, the fire and the individual trees exist independently of the epistemological lens used by that particular geographer as illustrated in Figure 2.1. Nevertheless, the epistemology or lens used to interpret the ontology (thing) has a profound effect on its interpretation. Every epistemological perspective imbues the observation with different meaning, and different ontologies come into view depending on the epistemology of the GIS user. Ontologies exist separately of the methods that humans use to study them, *but* they are interpreted through epistemology – or research perspective. Definitions of epistemology and ontology are a convenient way to distinguish between methods, agenda, and spatial entities, but the terms are frequently conflated simply because epistemology is closely

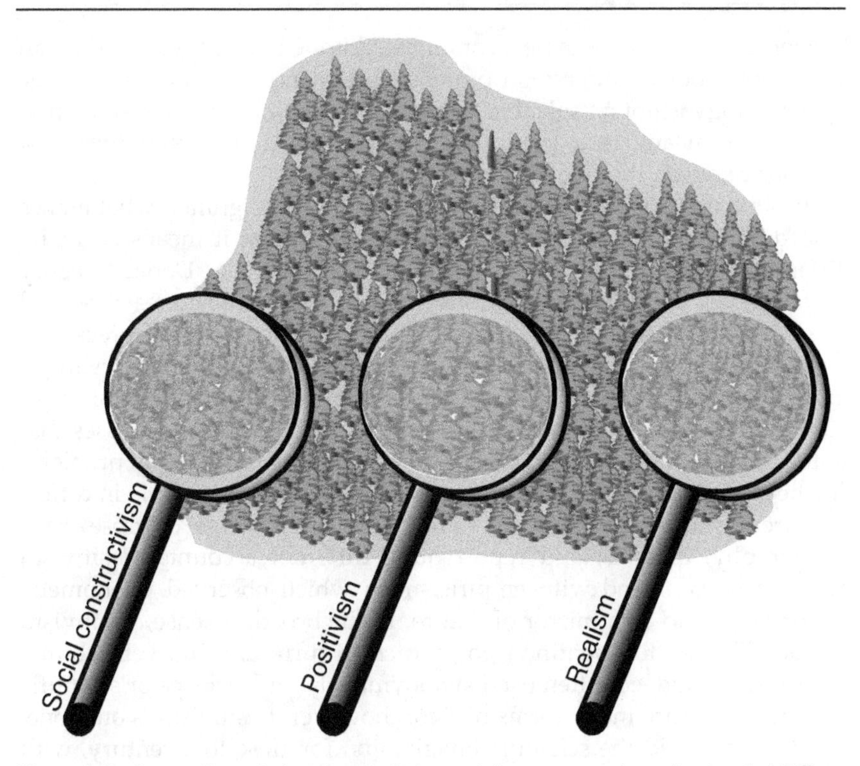

Figure 2.1 *Epistemology is the lens through which spatial phenomena are observed and studied. Different epistemologies such as positivism, realism, or social constructivism result in the apprehension of different entities or ontologies. (see http://www.blackwellpublishing.com/schuurman for color version.)*

linked to ontology. One's epistemology bears heavily on the resulting ontology. How one studies and understand the world contributes to the ontologies that are evident to the observer (as Figure 2.1 depicts).

Discerning ontologies is different again from representing them. *Representation* refers to the ways that phenomena are depicted after having been interpreted through an epistemological perspective. Representation can take the form of a picture, text, or, in the case of GIS, a map.

Epistemology

Of all the concerns raised by critics of GIS, criticism of epistemology was the most sensitive. Critics of GIS asserted that positivism was the epistemological basis of the technology's use and construction. GIS researchers were, in turn, defensive about accusations of positivism. Disagreements between practitioners of GIS and its critics over whether or not the technology is positivist have never been resolved.

Epistemology in GIS became a complicated issue because it hinged on an accusation poorly understood by researchers and ill-defined by critics. Epistemology is not a "solvable" point of contention but an examination of what it means is a preliminary step toward understanding past tensions between the two research communities.

"Positivism" has become something of a floating signifier; what makes it so difficult to refute when applied to GIS is that it means so many different things. Positivism is difficult to pigeonhole. Derek Gregory (1994b) notes that there are various versions of positivism, each related to a particular historical and philosophical context. A central precept of positivism, however, is that observation precedes theory. Observations must be repeatable and theories are constructed on the basis of those observations. One version of positivism, logical positivism, stresses that statements about the world must be verifiable. This is infinitely problematic because evidence is subject to perception. Different people in different circumstances can see the same forest fire, bridge collapse, rock stratigraphy, or soils, and report quite different accounts. Positivism can also be confused with empiricism in which observed phenomena are presumed to be a mirror of nature. In the broadest sense, positivism has been linked to scientific rigor, which in turn relies on verifiability. Positivism is, indeed, often used synonymously with science or scientific method. There are many forms of rigor, however. Positivism is only one, but it has secured the scientific imagination for close to a century, with the result that it is equated with reliable and repeatable results.

Scientific method traditionally refers to problem definition that relies on hypothesis testing. A null hypothesis that a relationship or fact is *not* true is posited. Then the scientists gather data, and conduct analysis in an attempt to prove the null hypothesis false. If successful, and other scientists accept the evidence, then the experiment can become a foundation for other hypotheses and theories. Science tends to be cumulative, building new hypotheses and experiments upon already proven *facts*. There are two obvious shortcomings to this well-entrenched approach to science. The first is that experiments can be misleading, and subsequent science can be built upon false assumptions. Examples of this abound, the long-standing belief that the sun revolves around the earth being one of the more striking. The second is that scientific methodology is particularly suited to limited data sets of the past. The development of scientific method was partly a response to financial and technical constraints of limited data availability. It allowed scientists to collect small amounts of data, and extend their limited observations to a wide range of phenomena. Situated historically, scientific method can be seen as a response to the high costs of data collection and laboratory analysis. Samples had to be limited and hypotheses rigorously constructed in order to produce results which had bearing on a wide range of situations. That period of limited data has been succeeded by data proliferation. Tools are available which generate informal hypotheses based on the data rather than preceding them. GIS operates in an unprecedented "data-rich" environment. Moreover, very little GIS research is conducted under the assumptions of positivist scientific method.

The question that emerges is why have critics focused so emphatically on positivism with reference to GIS. Part of the problem stems from a tradition of imposing a positivist framework on early GIS as a way of staging the discipline's entry into *science*. Ironically, the positivist label with which GISers were later saddled may have been partly elicited as a result of their own promotion of the technology as scientific. This reflects a cultural insistence on objectivity. A demand for scientific "objectivity" underlies the polity and legal basis of our society. There is a well-entrenched cultural insistence on record-keeping and formal language that is reflected in GIS architecture. Criteria for data recording are continuously reinforced institutionally by organizations such as the US National Spatial Data Infrastructure (NSDI) and the Open GIS Consortium that designate and monitor protocols for GIS interoperability. Neither GIS nor any iteration of it, whether democratically motivated or not, can eschew those criteria, short of extracting itself from the institutional structure in which it is embedded.

Ironically, despite charges of positivism, GIS researchers and scholars more frequently identify their epistemological slant as realist. Realism, like positivism, is difficult to define with precision. It can refer to forms of representation like sculpture or paintings that refer to the real world through the use of imagery. Philosophers define realism with more precision. Realism, in this context, implies that the world and events are linked to structures that cannot be clearly seen, but that can be discerned by studying the relationships between particular events. In science, realism refers to the abstractions that identify and explain causal structures for phenomena under very specific circumstances. A forest fire, for instance is an event that is linked to environmental relationships and social policies (structures) that give rise to fires. Realism presumes that there are facts (the structures that caused the fire) that are independent of the mind (that is, true or false in an absolute sense), but that can be discerned through study and observation (Sismondo, 1996). It is a kind of shorthand for theories that the evidence points to (or that the scientists believe that the evidence points to). Realism puts greater emphasis on specific conditions than does positivism. In that sense, it relies on a more circumscribed form of empiricism as it does not necessarily link causes (what caused this fire?) to other data. A realist would not link the structures that caused this forest fire to all other forest fires. Realism better accounts for the spatial-temporal location of entities by connecting them to specific situations. Realism is thus a more contingent epistemology.

There is a decided philosophical affinity with realism in the GIS community, despite a general agnosticism toward philosophical inquiry. And undeclared GIS users could be considered implicit realists by virtue of the fact that GIS is frequently concerned with prediction rather than explanation, and that requires identification of structural and causal mechanisms – both hallmarks of realism. Despite a philosophical affinity for realism among GIS researchers, an alternate argument can be made that the technology, and use of it, often approximates pragmatism!

Pragmatism is an approach to knowledge that incorporates changes as necessary to accommodate new evidence, or, more commonly in the case of GIS, technical difficulties. Pragmatism is antifoundationalist, tending instead to regard knowledge builders as participants rather than observers. Knowledge, in pragmatism, is instrumental but only as a tool for organizing the world (whether in digital or analogue environs). Truth is not absolute and can't be defined by epistemological criteria precisely because there is no outside position with which to discern it. Moreover truth can be revised. Pragmatism regards knowledge as derived from experience and scientific experimentation, and is skeptical about metaphysical rationales or posturing. GIS users, for instance, often fit the technology to the problem, and the two are developed in tandem. GIS data are collected based on the data tables and analytical capacities of the technology. Researchers rarely frame their investigative questions using hypotheses. Instead they provide proof by demonstration. These are all hallmarks of pragmatism. Moreover, GIS scholars tend to focus on local rather than generalized patterns. Questions like "where should we build the new light rail line" are not positivist, nor realist, but pragmatic.

Human geographers frequently viewed GIS as positivist, yet many GIS scholars identify as realist (though GIS researchers are as epistemologically varied as any group), and the technology has the markings of pragmatism. One can only conclude that GIS scholars work with a mixed epistemological toolkit that varies from positivism to pragmatism. This is not to eschew an epistemological examination of the technology but an attempt to problematize it. Phenomena can also be "read out" of GIS simply through the choice of data or classification scheme, and this carries implications for the construction and presentation of knowledge about the world. The limitations of GIS constitute an epistemology in themselves. David Mercer (1984), for instance, noted that we have elaborate analytical means to find the shortest distance to the hospital but fail to ask why so many people get sick. The trouble with geometric language used in GIS is that it frequently fails to explain. It may be that these types of limitations are what critics were trying to describe by using the shorthand of "positivism."

At the end of the day, it may be difficult to discuss GIS, the computational technology, in epistemological terms at all. It doesn't matter whether your goal is to locate a freeway or to create a socially responsible GIS, ultimately it has to be computationally viable. Human geographers, however, often uses discourses, or vocabularies, which have little overlap with the mathematical reductions of GIS. Their vocabularies are simply quite different. At the computational level, GIS is implemented using numbers and codification. Geometry, algebraic descriptions of spatial relations and logical concepts like union, intersection, inclusion are the currency of GIS. In this context, Michael Goodchild has raised the question of whether it is even possible to move epistemological debate to a technical level (Schuurman, 1999a). There may be a basic incompatibility between the generality of epistemological discussion and the tools of computation in GIS. There is a discord among discourses. It may be that epistemology is too abstract a notion to "map" onto GIS, that we can only talk about applications and implementation of GIS or about error and omittance in data – pragmatism, in effect. In this vein, it is useful to examine the building blocks of GIS – its ontologies – in order to better understand the foundational concepts that evade epistemological categorization.

Ontology

The short definition of ontology offered above is potentially misleading as it refers to what philosophers and social scientists mean when they use the term *ontology*. To a philosopher, the word implies the essence of being, an ultimate and stable reality. Computer scientists use ontology in a different way – to mean a formally defined set of objects in which all the potential relationships between the objects are also well defined. The earth's surface, for instance, abounds with spatial objects (they are referred to as spatial entities when represented in GIS) such as forests, roads, houses, bridges, telecommunications towers, and urban shopping malls. To represent these objects in a GIS requires a method of *encoding* them digitally as entities, and then encoding the relationships between them. A road can cross a bridge, but a river must run under it, and bus stops occur only along bus routes. Rules or relationships between spatial entities are just as important as their definition. This closed, formal universe is referred to as an ontology in computer science.

The problem is that GIS share intellectual territory with both human geography and social science as well as computing science. As a result, the notion of ontology becomes more vague – attenuated by competing influences. GIS researchers use the tools of computer science to create the ontologies, but the concepts of philosophy to refer to the spatial entities manufactured by different methods of encoding. There is recognition in GIS that different *data models* produce dissimilar ontologies for what are the same objects on the ground. Thus the way that data are structured inside the computer has a profound impact on how the entities appear on the screen, and, more importantly, the results of analysis. Figure 2.2 illustrates how data models mediate the representation of spatial entities in GIS. The pictures in the figure represent the entities that must be represented. In order to depict them in the computer, they must be interpreted through data models, and then represented through map symbology.

Data models reflect different ways of seeing the world. They are ways of imagining space – in order to render it in a computational environment. Data models suggest highly formalized inscriptions, scrawled on a blackboard, describing the nature of space. Their implementation in GIS has been much more prosaic – tied to "0"s and "1"s and computer architecture. When Roger Tomlinson and Lee Pratt participated in the development of the CGIS (Canadian Geographical Information System) in the early 1960s, they were concerned with how to instill the land use characteristics of Canada into a digital environment. The nature of geographical space or how to represent it didn't worry them. The vector-

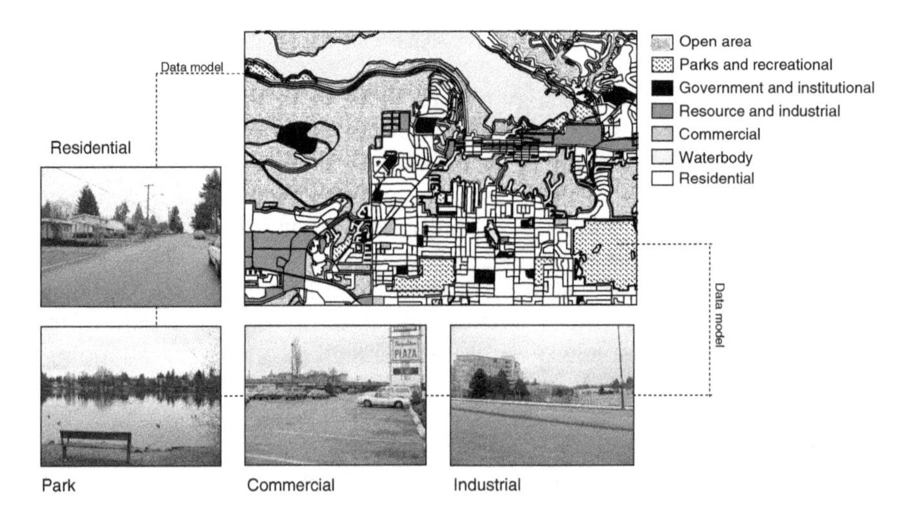

Figure 2.2 *Data models are used to represent spatial entities.*
Data models allow only some facets of the real world to be represented. These are then re-represented through map symbols in order to portray them to the user. There are several levels of mediation between the real world and the map.

based system that they developed in the process was a product of exigency and a cartographic convention based on using points and lines to depict spatial objects. Subsequent systems were largely raster (grid) based until the 1970s when the Harvard Graphics Lab began to develop more sophisticated vector representation. In the 1990s, GIS began to follow computer science in implementing an object-oriented programming approach in which classes and instances of geographic objects are hierarchically defined like the genealogy in family trees. These three approaches to modeling space constitute the current selection in data models. Each divides and defines space, and relations within it, differently. All were, to various degrees, feats of engineering rather than philosophical triumph though arguments for the ontological fidelity of each has subsequently been made.

The vector data model is the most ubiquitous in GIS, and most closely resembles traditional maps. Vector data models are constructed from points, lines, and areas. The *point* is the zero-dimensional primitive; lines are considered one-dimensional and are constructed by an arc or chain linking two points. Areas are defined by sets of lines, surfaces are areas that include height or another dimension such as population density that can be used to portray relative elevation. In a vector data model, polygons are synonymous with areas. Three-dimensional surfaces are, likewise, built from areas. The relationship between points, lines, and areas is illustrated in Figure 2.3 (Schuurman, 1999b). Vector-based GIS have traditionally been distinguished from cartographic points and lines by their data structures which include topological information, based on adjacency and connectivity. Topological information allows redrawing of areas without drawing points or lines twice because it tells the computer which polygons share which lines, and which areas are above, below, or to the side of another area. More importantly, it is used to expedite the computation of spatial queries such as "how many counties are contiguous with the water reservoir? Or which logging roads are within 500 meters of the White Pine forest?"

Raster data models divide the world into a sequence of identical, discrete entities, by imposing a regular grid. Frequently the grid is square. Figure 2.4 illustrates the raster view of the world in which the characteristics or attributes of each grid cell (or raster) are linked to that specific location. Usually a series of attributes, such as population density, soil type, or presence/absence of a fire hydrant are linked to a raster coverage. There may be hundreds of attribute layers for one geographical area, but each attribute layer includes the same grid cells, and each cell contains a single value for each attribute. Raster data models are distinguished by their conceptual simplicity and ease of implementation. Raster systems are widely used in applications which employ remotely

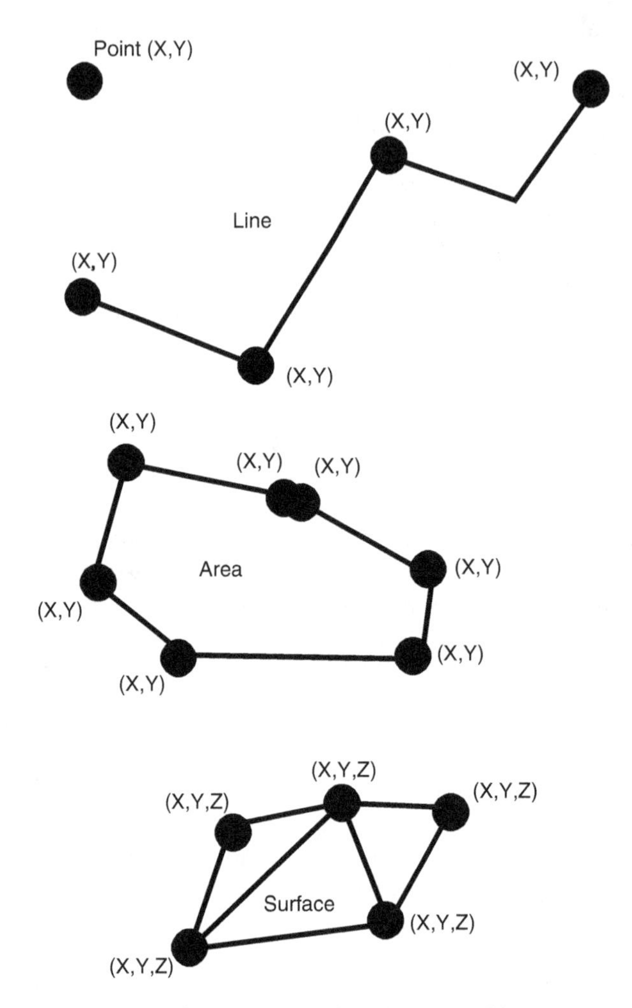

Figure 2.3 *Points are the basis for representation of spatial entities in GIS. Lines link points, and areas are based on lines. Surfaces are built from areas. This simple set of building blocks is a robust method of mimicking traditional cartographic representation.*

sensed images as satellite imagery is represented using a gridded network. Raster data models are well-suited to operations which determine travel time or direction of water drainage as the grid system allows easy comparisons of values between neighboring cells. Figure 2.5 illustrates the path of a ball as it rolls down a slope. The ball, like water or snow on a

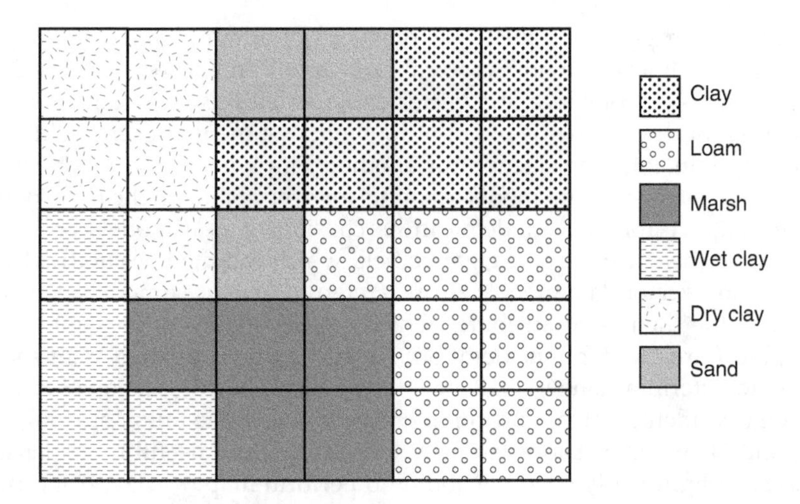

Figure 2.4 *In the raster view, the earth is divided by a regular tessellation. Each grid cell represents a location and the resolution of the coverage is determined by the grid size. Attributes are assigned to each cell, one attribute per layer. Location is everything.*

Source: *Figure adapted from Schuurman, N. (1999b) with permission.*

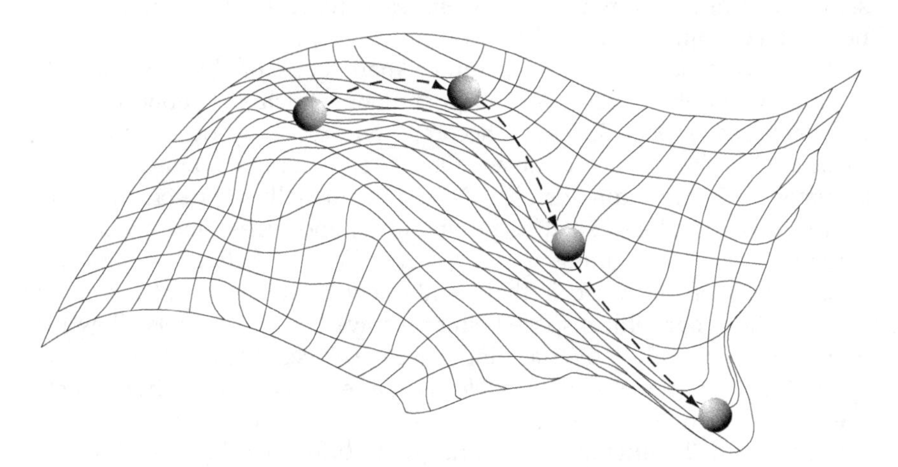

Figure 2.5 *The least-cost path is the one that requires the least energy for the ball to travel.*

hillside will choose the course of least resistance – or lowest elevation. These examples are based on raster grid cells. A least-cost path can also be defined for a vector data structure to find, for instance, the fastest automobile travel route. Raster and vector data models often share the same functionality, but it is achieved through different methods.

Vector data models are predominant across a spectrum of applications, from corporate to public to engineering. The polygons which comprise a vector landscape are frequently expressions of administrative and legal fiat. Census tracts, provinces, postal code zones and school districts are typical divisions of space in a vector data model. It is important to note that each of these systems of apportionment assume that every point on the map falls into one category or another. Empty space does not exist. Likewise, the raster data model accounts for every point on the map. Both raster and vector data models can be considered *field* models because they treat the coverage as a gigantic field which can then be subdivided. Raster and vector data models are also referred to as *layer* models. Layers refer to the themes or attributes which are registered to the same geographical area under consideration. Layers can be compared with each other to determine a new set of attributes such as the intersection of lower income residential areas and commercial zoning or the union of Douglas Fir and Hemlock in a forested area. One of the most useful functions of GIS has traditionally been the ability to perform map overlay, using the principles of Boolean algebra, to derive new attributes from a combination of layers. This has given rise to the characterization of GIS as a "layer-cake" view of the world.

Object data models are an alternative to the vision of the world as a series of locationally registered layers, each representing a single attribute. Objects came into GIS from computer science in the late 1980s and early 1990s (Goodchild, 1995). Rather than focus on location, object-oriented GIS defines geographical phenomena, such as telephone poles or streets, as objects. Location becomes one of many other attributes associated with a particular object. Objects can be points, lines, areas, or volumes with three dimensions. Objects are usually represented using vector building blocks such as points, lines, and areas, and are a very different way of imagining space. A key difference between object and field models of the world is that, while field models account for every point in the geographical space being portrayed, object models allow for empty space as well as overlapping objects – objects occupying the same space. Fields are concerned with *what* is at a certain location while objects are concerned with the object itself, and location is a secondary characteristic. Figure 2.6 illustrates the conceptual differences between the two data models.

Every object in an object-oriented GIS data model is not defined individually. That would be a painstaking task. Rather, groups of like objects are organized into classes. Vehicles used for transportation might comprise one class. Subclasses would include cars, trains, trucks etc. Each class and subclass have attributes which apply to the entire group. In addition, operations or methods that describe possible pertinent actions, can be defined with reference to the class. Attributes and procedures are bequeathed through a hierarchical system of inheritance as illustrated in

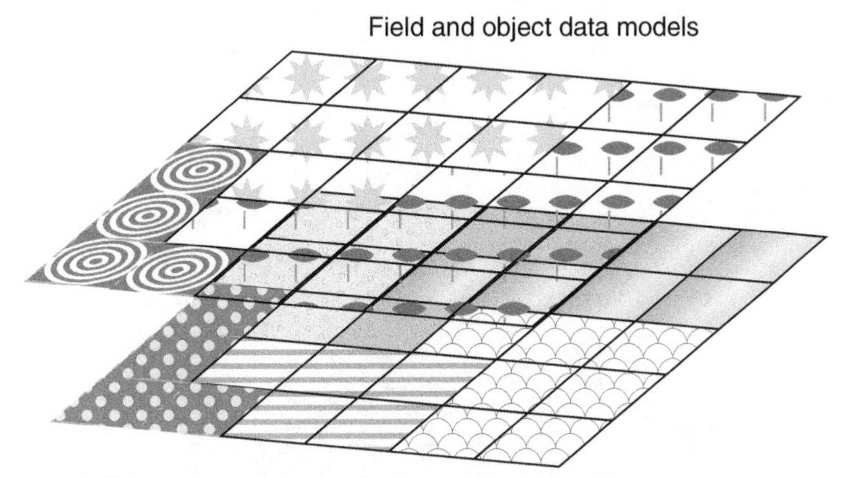

Field data models can be envisaged as layers which register to the same geographical coordinates, each containing information about one attribute or theme.

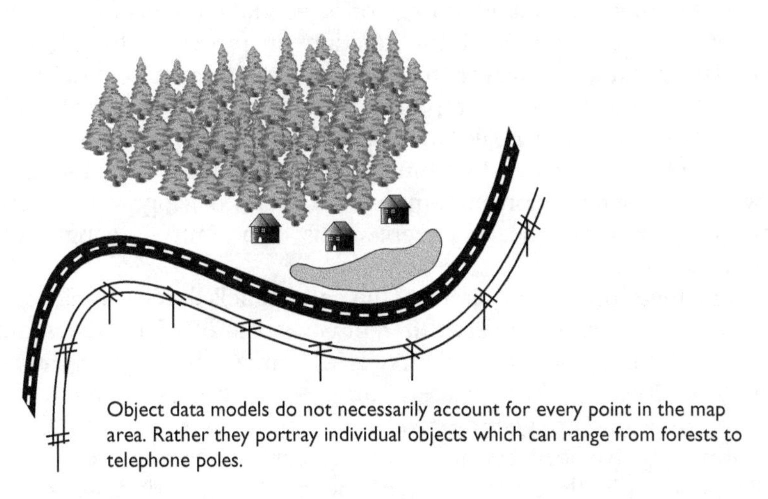

Figure 2.6 *Conceptual differences between field and object data models.*

Source: *Figure adapted from Schuurman, N. (1999b) with permission.*

Figure 2.7. Each parent class contains all attributes that are common to its subclasses. Using a hierarchical system allows rapid updating of general characteristics. That is not, however, the chief conceptual attraction of object-oriented data models. Rather they are believed, by some, to more closely parallel human conceptualizations than location-based field

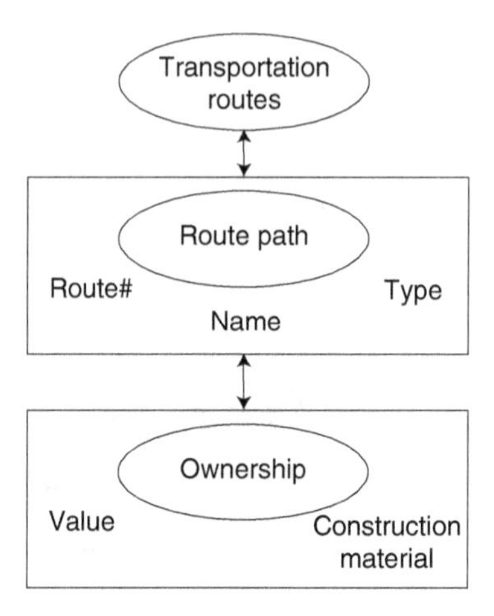

Figure 2.7 *Object-orientated inheritance.*

models. Another reputed advantage of object data models over fields is that entities can be defined by function or process rather than by name. This assumes, however that functions can be precisely delimited. These, like many other assumptions about data models, merit critical inquiry. Both field and object models rely ultimately on a view of the world in which neutral and absolute space is assumed. Nor does either allow the characterization of complex, interrelated geographic entities. They are each simplistic characterizations of a complex geographical reality.

Despite their inherent limitations, the data models that we are forced to choose between represent very different ways of seeing and representing the spatial world. Object data models epitomize the idea of discrete, separate entities in a frictionless neutral space. Both field models are ways of dividing up geographical space into discrete units. Rasters are intuitively friendly as each location corresponds to an attribute. Raster also implicitly portray one of the most basic principles of geography – spatial autocorrelation – or the tendency of like phenomena to be located close together. This has enabled the development of a number of spatial analytical operations that exploit this spatial property. Another reason that rasters remain so prevalent is that satellite technology provides us with mammoth amounts of data in this form. Vector data models, however, frequently correspond to human divisions of the landscape (Couclelis, 1992). Jurisdiction dictates the division of space, and attributes are positioned within political or administrative boundaries; it is a manager's view of the world. For many applications, routine vector division of space makes imminent sense. It behooves the analyst using census data to divide the world into census tracts. It makes less sense to use census tracts or provincial boundaries to track geological strata or animal movement. But researchers who inherit a vector-based data set often do just that. Users assume greater ontological flexibility than is justified by the data model precisely because they disregard epistemology.

Each of these data models is inherently limited. Both fields and objects are reductionist. They are products of the presumption that the best way to solve a problem is to make it into smaller problems and use the subsolutions to make a total solution. Data models and the queries that they enable are based on the Aristotelian logic that underlies reductionism. Three laws form the basis of Aristotelian logic. The first is the law of identity which states that everything is what it is. A house is a house. A car is a car. The second is the law of noncontradiction which states that something and its inverse or negation cannot both be true. The third is the principle of the excluded middle which holds that every statement is either true or false. No semitruths or partial membership in a set. Spatial phenomena, however, seldom obey these laws. Geographical space is crowded with fuzzy borders such as those between mountains and valleys, wetlands and marshes, and urban and rural. Moreover, spatial entities change form over time, and depending on context. Ambiguities such as these are difficult to encode in either fields or objects. Furthermore, fields and objects ultimately rely on an idealized vision of empty space, once they are geocoded. In both instances, a neutral, pliant space is assumed. Despite clear evidence that many geographical phenomena do not obey these premises, they structure all examinations of ontologies, simply by virtue of informing the only data models we have. Attempts to grapple with ontologies at a higher theoretical plane have led inexorably back to reputed virtues of objects or fields.

For the moment, objects and fields are the only way that GIS users have to represent spatial objects as entities in a digital environment. Objects and fields define the ontological possibilities of GIS. Choosing a data model is therefore an important decision in terms of representation and analysis. Early GIS researchers harbored the illusion that data models would be chosen based on the structure of the phenomenon itself, that they would be painstakingly developed to truly "reflect" spatial events. This approach was first articulated by the French mathematical geologist, Francois Bouillé at a 1977 Harvard meeting on topological data structures (one of the first and most famous GIS conferences) (Dutton, 1977). The idea was that different domains (with different epistemologies) would lead to different data models. Limitations of this assumption became evident immediately. Roger Tomlinson (1984) was quick to point out that any institution, after having commissioned, at great cost, an appropriate system for its spatial analysis tasks, would then declare: "Ah, we've got the system. We must now use it for everything." Indeed, the GIS market has resolved around two fundamental data models.

One might conclude that GIS is thus bankrupt from an ontological perspective, but this is a mistake. Every system of representation is fundamentally limited. The English alphabet, for example, comprises only 26 letters that, in combination, are often assumed to be able to communicate all concepts to all audiences. It is well known, however, that many ideas, feelings, and intuitions are not well advanced through written text. GIS, likewise, works within a constrained system of representation (Schuurman, 1999b). Furthermore, there is no perfect model. Models of the earth and its processes are inherently constrained. Thus the famous quip, "all models are wrong, some are useful." The possibilities for communication are nevertheless profound. Dismissing the power of GIS on the basis of either epistemology or ontology fails to account for the poverty and potential of all systems of representation. It may be more productive to examine how representation in GIS is developed and used. This requires looking to contemporary social scientists and human geographers for the intellectual tools to understand how social processes are written into technology.

Looking for the Social in GIS: Using Precepts from Human Geography to understand GIS

Critics of GIS from human geography were successful in pointing to the importance of understanding the epistemological and ontological repercussions of different technological directions in GIS. Indeed, there is a burgeoning parallel pursuit in the GIS research community to extend the ontological integrity of data models, and to create systems of classification that exhibit greater ontological flexibility. Many of these efforts have been framed in terms of both epistemology and ontology. Human geographers have meanwhile drawn attention to ways that technology is influenced by social processes. Recognition that science is part and parcel of our cultural fabric, rather than an objective lens on reality, has permeated the social sciences over the past quarter century. The idea that scientific direction might be influenced by constraints such as funding directives and cultural views of gender does not seem that radical today. It is widely accepted, for instance, that clinical trials on health issues have largely excluded women, and that late-nineteenth-century research based on skull measurements was racist. But until about thirty years ago, the History of Science and the Philosophy of Science were fields that focused on the great names of science, and justifying their "discoveries" without thinking critically about the process of science nor its goals.

Ernst Mach, a philosopher of science first challenged these disciplines in the beginning of the twentieth century by arguing that science should be designed to help society (Ziauddin, 2000). He was opposed by Max Planck who was an advocate of an autonomous science which was independent of society. Planck's arguments were successful over Mach, and entrenched a Platonic ideal of the scientist working selflessly in *his* laboratory, with the sole motivation of unearthing the "truth." Interestingly, Planck was one of the scientists who founded the Vienna Circle which had promoted a strong form of scientific positivism (Ziauddin, 2000). World War II and the deciding role that scientists played in partisan warfare was a factor in dislodging the noble, rational and disinterested view of science. The development of a substantive critique of science and technology began about the same time.

It arguably began with Heidegger's famous essay *The Question Concerning Technology* (1982). Heidegger argued that the process of experimentation in science is inherently "theory-laden" as it imposes a framework on the earth. This speaks to concerns that human geographers have had with respect to epistemologies associated with GIS. Heidegger argued moreover that science and nature are configured in ways that are familiar to us in the same way that Wittgenstein argued that language grows to reflect the structure of the world – as humans perceive it (Heidegger, 1982). These founding concepts slowly made their way into social science over the course of the twentieth century, and the research domain of *Science and Technology Studies* (STS) resulted. STS is a broad rubric for many researchers who examine ways that science and society are intertwined.

It was Thomas Kuhn's famous book *The Structure of Scientific Revolutions* that most famously upset notions that science is an a-philosophical, culturally independent pursuit (Kuhn, 1970). Kuhn illustrated that scientists solve problems within accepted formulas or belief systems that are periodically overthrown in favor of new paradigms. The implication is that scientific findings are tied to established paradigms, and contingent on the frames of reference within which they are discerned and disseminated. It is interesting to note that Kuhnian accounts of scientific paradigm shifts dealt exclusively with conceptual science. GIS is, by contrast, "tool science" (Dyson, 1999; Schuurman, 1999b). We tend to think of science and technology as purveyors of stand-alone concepts but, since World War II, both have increasingly been constrained by tools. Today, almost all scientific and technical production is a joint production between machines, computers, measuring devices, and researchers. Goals are defined by the possibilities of the technology and are reshaped according to its restrictions. Indeed, tools have driven a great majority of recent scientific "discoveries" Crick and Watson's identification of the double helix structure of DNA was determined as much by tool as concept (Watson, 1969). Without X-ray crystallography, they wouldn't have envisaged the endeavor nor succeeded (Dyson, 1999). Tools likewise created GIS and GIScience. It is partly this intermingling between machines and science that drew the attention of social science researchers, and fostered the rapid growth of STS in the latter part of the twentieth century.

The roots of present day STS are perhaps most famously in the Sociology of Scientific Knowledge (SSK) Strong Programme developed in Edinburgh in the 1970s. There sociologists sought to demonstrate that you can study scientific knowledge using sociology. Among the tenets of SSK is that knowledge discovery is affected by social and economic conditions. Thus a discovery such as penicillin is not an inevitable result of a relentless pursuit of pure knowledge; it does not reflect the world per se. Rather such discoveries require a certain set of external conditions such as a receptive audience, demonstrated use, and a network through which to disseminate the "discovery." One trouble with the Strong Programme is that, like science, it assumes a single unassailable reality, some that social scientists are now less likely to take for granted. Feminist perspectives on science and technology have, for instance, demonstrated that there is frequently an androcentric bias to scientific investigations. Sardar Ziauddin (2000) gives the example of traditional scientific explanation of men's role as hunter-gatherer, and women's of child-rearer. These were based on interpretations of "chipped stones" that presumably demonstrated early male use of weapons. A different interpretation might yield that the same chipped stones are tools that women used to kill animals, eviscerate corpses, and dig up roots.

The doctrine of SSK has been modified in the intervening decades by people who recognized that science is culturally constructed but found reason to believe that at the bottom of it all, a real world exists. This branch of STS, characterized as cultural studies of science, is based on the work of Bruno Latour, Donna Haraway and Joseph Rouse, among others (Haraway, 1991; Latour, 1987; Rouse, 1996). They acknowledge that there may be a "real world" but maintained that the way that we know and represent it is socially constructed (Schuurman, 1999b).

The most famous (in geography anyway) strain of STS is Actor Network Theory (ANT), and is based on the idea that scientific entities come into existence through social processes, especially their interrelationships with humans and other nonhuman agents (such as measurement devices and computers) (Schuurman, 2000). ANT was popularized by French anthropologist, Bruno Latour (1988; 1999), who demonstrated that scientific discoveries and technological products as diverse as penicillin and rapid transit systems were developed through networks of humans and infrastructure. Changing the network components influences outcome of the science and technology. Andrew Pickering (1995), a physicist and STS researcher, has written that he believes that quarks are not the inevitable product of scientific discovery; their detection and naming depended on a host of contingent factors that, if altered, might have resulted in a different entity with different characteristics. This is not to undermine modern physics, but to recognize that the building blocks of nature can be differently classified, and differently understood depending on epistemology and underlying assumptions. Splinter groups of social scientists studying science represent a far greater range of epistemological positions than those mentioned here. They all share, however, a conviction that science and culture are inseparable. The direction of science and technology can change at any point, and influence resulting developments. The message that we can take with reference to GIS is that you can understand a lot about a technology by examining the social forces that acted on its development.

Cultural Influences on the Development of Theory and Technology in GIS: The Example of Generalization

Technology is not an inexorable force that societies must adapt to like weather patterns or earthquakes. Rather, technology is developed for social purposes, and in tandem with social goals. Often the development of technology is constrained by traditional ways of thinking about problems. Indeed, it is impossible to separate scientific truth, and its technological production from the social parameters of their inception. GIS theory, for example, is influenced *both* by cultural and technical factors. There is a social dimension to every digital decision, just as there are real technical strictures to conceptual formulations. Intellectual and cultural influences as well as technical influences are interlocked at every stage. This is evident in the development of generalization theory in GIS (Schuurman, 1999b).

Generalization is the elimination of map detail as scale decreases. A map of a city block at a large scale, say 1:2,500, might depict individual buildings, including sidewalks and alleyways. As scale decreases, geographic objects must be eliminated. At 1:10,000, sidewalks would disappear from the map; at 1:500,000 only a colored outline indicating an urban area transected by roads and freeways would remain. At 1:2,000,000 a dot indicating a city would suffice. Figure 2.8 illustrates how an urban area might be depicted at different scales. As the scale decreases, the level

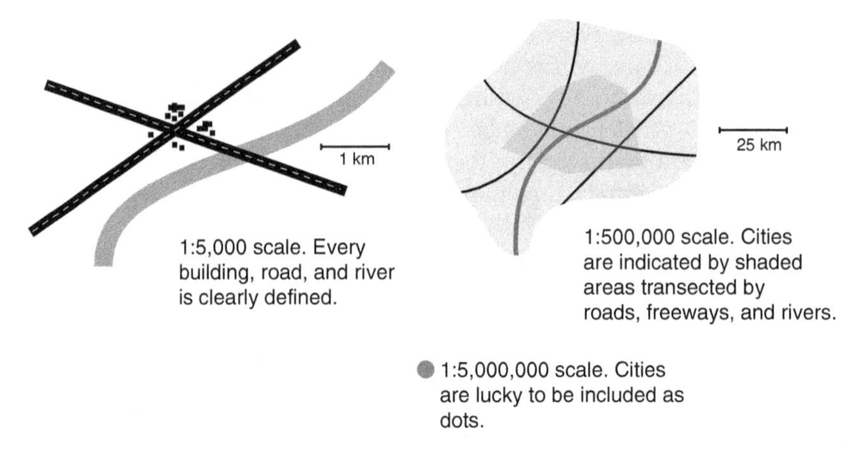

Figure 2.8 *Characteristic map symbology for cities at different scales.*

of detail also decreases. Reduction of detail – analogous to zooming out – is effected by different mechanisms including aggregation of features, simplification of symbology, and removal of spatial entities. Generalization is also context sensitive; whether a geographic object should be displayed, depends not so much on its size but its value both in the context of the map theme and its surrounding features. A large local shopping centre is unlikely to be displayed even on a large scale map if the map theme is precipitation and a number of isolines converge at that point. However, if nothing much is taking up that space, the shopping centre might be included. There is an element of aesthetic judgment in manual cartography, but GIS researchers have attempted, over the past 25 years to automate the process. There is a corresponding body of generalization theory which describes the possibilities of digitally encoding intelligent map simplification as scale decreases (Schuurman, 1999b).

Some transitions between scale are well worked out in terms of displaying individual objects. Transportation routes, waterways, and settlements are associated with specific symbols. Rules for their representation at various scales are established and their symbology is not subject to interpretation; this makes the computerization of generalization easy to encode. The main *impediment* to generalization is not representation but conflict between features and the need to incorporate contextual information about geographical objects. For instance, given that bird habitat and wetlands are inextricably linked, generalization should not include one while excluding the other. Information about linkages and associations of meaning between geographical objects cannot be extruded to the map unless they are first encoded in the data base.

In addition, many phenomena and relationships represented on maps are difficult to transform between scales. Traditionally, cartographers made generalization decisions individually taking into account the map theme, relative importance of features in question, context, classification system, adjacency priorities, form, and balance. Computerization resulted in a need for algorithmic solutions to processes which were never explicitly rule-bound though they were subject to convention. Generalization theory is, in effect, the search for computerized techniques to do something that was never consistently executed or theorized before the advent of GIS (Schuurman, 1999b).

Generalization theory began to evolve in the 1970s, most famously with the Douglas–Peucker line generalization algorithm. Tom Poiker (he later changed the spelling of his name) and David Douglas were trying to reduce the number of points used to describe a line, not so much because they wanted to use simplified – or generalized – lines in GIS, but because the cost of hard drive storage space was astronomical at that time. Their efforts to reduce the number of points in a line ironically became the basis for generalization of line features for two decades. They developed their line generalization algorithm during a period in which mass digitizing of existing maps was being conducted in order to populate spatial databases. This was also an era in which many geographical features were symbolized using lines. By reducing the number of (x,y) points associated with lines, they could significantly decrease the storage space associated with vector lines and polygons – which are built from lines. The technique that they developed did not address cartographic concerns about the quality of GIS output, but did accomplish the task at hand. Figure 2.9 explains how the Douglas–Peucker algorithm works, and illustrates its elegant simplicity. The astounding tenacity of the Douglas–Peucker line generalization algorithm – it is still resident in ArcInfo® 8 – speaks to the longevity of technology once ensconced.

The best method of generalizing the line from a representational point of view was not achieved, nor was the algorithm the most computationally efficient, but it did achieve the stated objectives of reducing hard-drive space, and thus cost. This was a clear example in which economics dictated the development of a technology. There is no longer an imperative to minimize hard drives space – it is now very inexpensive – but the algorithm remains with us. This speaks to the role of contingency in shaping future technologies.

Solutions to technical problems are also circumscribed by dominant beliefs or intellectual premises. Generalization, for instance, provides an example of how technical products are shaped by prevailing ways of

The Douglas-Peucker line generalization algorithm
(adapted from Chrisman, 1997)

0. Original line (to be generalized).

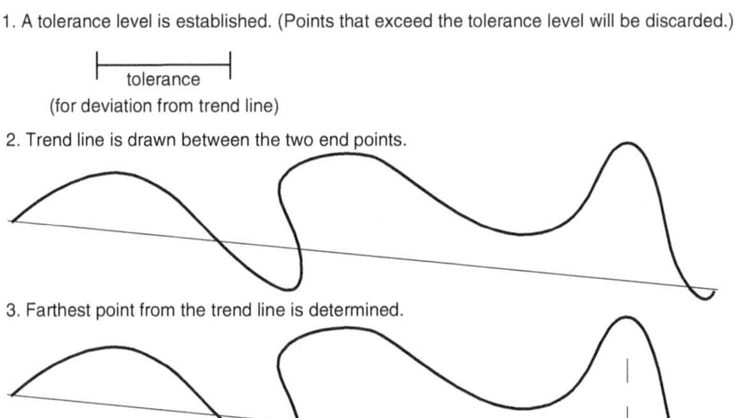

4. Distance to point is compared to the predetermined tolerance. (If smaller, then reject point, else accept point.)

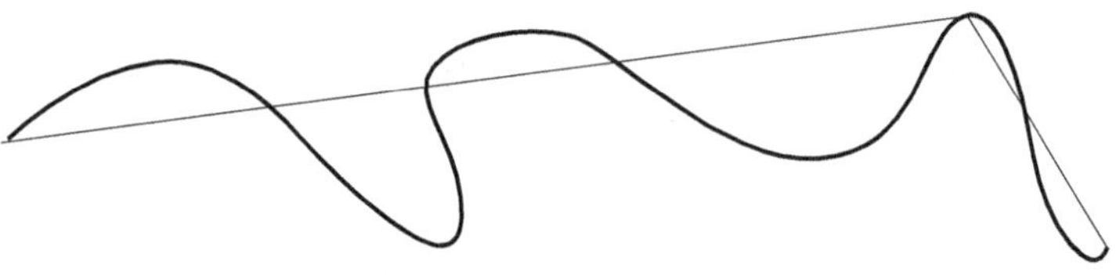

5. Whenever a point is accepted, two new trend lines are drawn, as in the example above.

Figure 2.9 *The Douglas–Peucker line generalization algorithm is the most popular of its genre. First published in 1973, it continues to be the workhorse of generalization. It is simple and elegant, but ironically, its original purpose was data reduction rather than generalization.*

Source: *Figure adapted from Schuurman, N. (1999b) with permission.*

thinking about a problem. This is evident in the way that computerized generalization was very slow to address the distinction between model and cartographic (or representational) simplification. "Model," in this case, refers to the database and "cartographic" to its display. In order to appreciate the difference between model and cartographic aspects of generalization, it is necessary to think of the map one sees displayed on the screen as the tip of an iceberg – as illustrated in Figure 2.10. The body of the iceberg, lying hidden below the surface, is the database. In order to simplify the map, it is necessary to simplify the database from which it is produced. Model-based generalization focuses on reducing the detail in the database while cartographic generalization simplifies the display.

In traditional cartography, there was little differentiation between data and map simplification. They were so closely tied as to be synonymous;

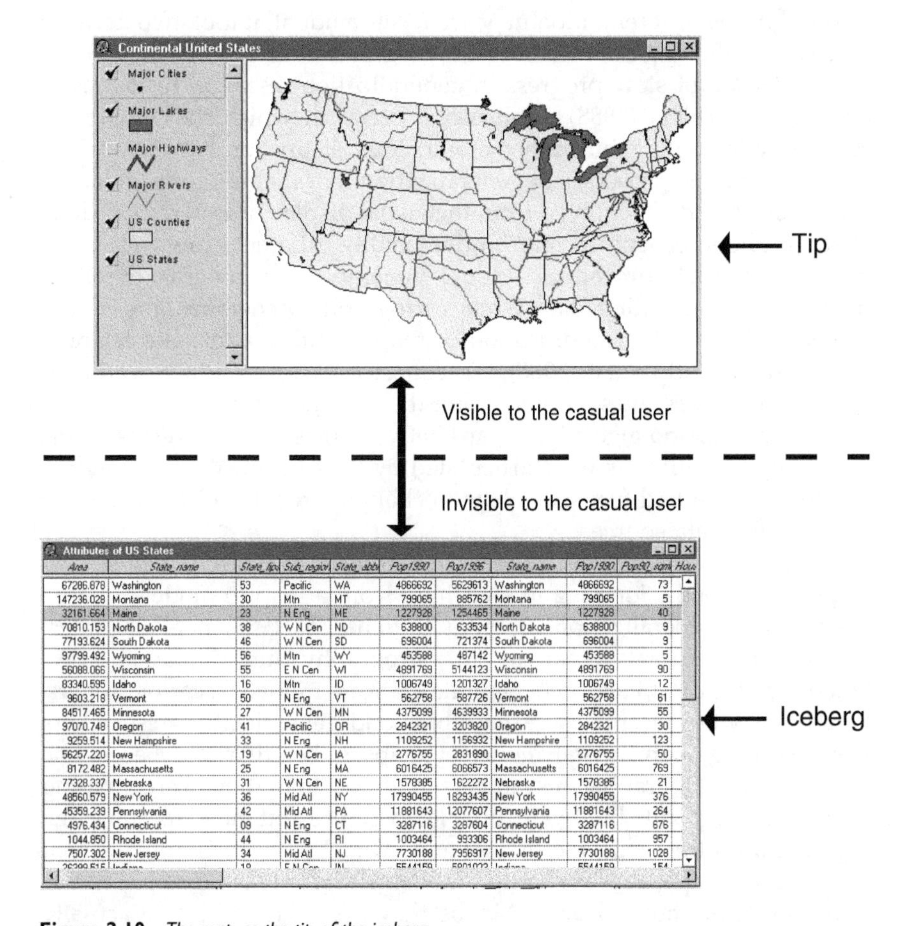

Area	State_name	State_fip	Sub_region	State_abb	Pop1990	Pop1996	State_name	Pop1990	Pop90_sqmi	Hou
67286.878	Washington	53	Pacific	WA	4866692	5629613	Washington	4866692	73	
147236.028	Montana	30	Mtn	MT	799065	885762	Montana	799065	5	
32161.664	Maine	23	N Eng	ME	1227928	1254465	Maine	1227928	40	
70810.153	North Dakota	38	W N Cen	ND	638800	633534	North Dakota	638800	9	
77193.624	South Dakota	46	W N Cen	SD	696004	721374	South Dakota	696004	9	
97799.492	Wyoming	56	Mtn	WY	453588	487142	Wyoming	453588	5	
56088.066	Wisconsin	55	E N Cen	WI	4891769	5144123	Wisconsin	4891769	90	
83340.595	Idaho	16	Mtn	ID	1006749	1201327	Idaho	1006749	12	
9603.218	Vermont	50	N Eng	VT	562758	587726	Vermont	562758	61	
84517.465	Minnesota	27	W N Cen	MN	4375099	4639933	Minnesota	4375099	55	
97070.748	Oregon	41	Pacific	OR	2842321	3203820	Oregon	2842321	30	
9259.514	New Hampshire	33	N Eng	NH	1109252	1156932	New Hampshire	1109252	123	
56257.220	Iowa	19	W N Cen	IA	2776755	2831890	Iowa	2776755	50	
8172.482	Massachusetts	25	N Eng	MA	6016425	6066573	Massachusetts	6016425	769	
77328.337	Nebraska	31	W N Cen	NE	1578385	1622272	Nebraska	1578385	21	
48560.579	New York	36	Mid Atl	NY	17990455	18293435	New York	17990455	376	
45359.239	Pennsylvania	42	Mid Atl	PA	11881643	12077607	Pennsylvania	11881643	264	
4976.434	Connecticut	09	N Eng	CT	3287116	3287604	Connecticut	3287116	676	
1044.850	Rhode Island	44	N Eng	RI	1003464	993306	Rhode Island	1003464	957	
7507.302	New Jersey	34	Mid Atl	NJ	7730188	7956917	New Jersey	7730188	1028	

Figure 2.10 *The map as the tip of the iceberg.*
Maps have traditionally been thought of as a model of reality. In GIS, however, the map is an ephemeral by-product of computational analysis. Hidden to the viewer, the database and algorithms are the motor behind the map product. In this scenario, the map is the tip of the iceberg.

Source: *Figure adapted from Schuurman, N. (1999b) with permission.*

the map was the singular repository for the data. Granted, cartographers worked from data sources, but when the map was produced, the sources were stored away in files. These files were quite separate from the map, and certainly not accessible to the map users. By contrast, generalization in GIS involves simplification of the database followed by appropriate revisualization of the information. These two steps are distinct yet inseparable. The data are never disassociated from the map, but the principles of generalization are very different for both. This model/cartographic distinction is fundamental to computerized generalization (Buttenfield, 1991; Schuurman, 1999b).

Though this differentiation may seem self-evident, it took two decades of generalization research for it to be recognized and articulated. In 1988, after a decade of slow progress in generalization research, Kurt Brassel and Robert Weibel (1988) published a paper in which they made the seemingly innocuous but rather momentous distinction between model and cartographic processes of generalization. For the first time, database processes were differentiated from the visual display of generalized data. Cartographers recognized that fundamentally different aims were associated with each procedure. Model generalization consists of either filtering or aggregating data while cartographic generalization is concerned only with the modification of maps entities, and their arrangement (Brassel and Weibel, 1988). It is concerned with avoiding overlap of roads and railway lines, and ensuring that the map is not overcrowded. The conceptual and algorithmic gaps between these two processes seems apparent but until they were articulated by Brassel and Weibel, generalization research had focused solely on changing display elements, while ignoring the data source.

A culture of cartography, especially dominant in North America, was responsible for the long lag in switching from a map to model-oriented approach to generalization. Barbara Buttenfield (1988, personal interview), a preeminent researcher in cartography and digital libraries, noted that many GIS researchers in the US were trained as cartographers. Paradigms were essentially cartographic and generalization researchers were working with mental model of maps rather than databases. She recalls that cartographic researchers realized that they were missing an important point, but were unable to figure out how to solve the problem until they read the Brassel and Weibel paper: "The reason that Kurt and Rob's paper was so important was that we read that paper and said they know how to do it. Maybe they don't know how to actually implement it but they at least intellectually can get around the cartographic impasse" (Buttenfield, 1998, personal interview; Schuurman, 1999b). Europeans had developed a new paradigm based on a separation between data and display. This critical distinction between the database and the map enabled the generalization community to bridge an important hurdle.

Two patterns of developing GIScience theory and resolving technical problems emerge. First, economic pressures bear on the way that solutions to problems are resolved. The Douglas–Peucker algorithm for line generalization was developed to simplify lines, but it was also designed to eliminate superfluous points that were expensive to store in the 1970s when hard drive space was at a premium. Second, the ways that a problem has traditionally been envisaged has bearing on how GIScientists work to resolve the problem. In the case of generalization, centuries of cartographic tradition led to a focus on the map as the sole object of simplification. In digital cartography, however, it is the database that generates information and map objects, and, in many cases, the database is the starting point for inclusion of more or less map detail. These two insights enable us to see the GIScience and other technologies as responsive to intellectual (and cultural) as well as technical processes.

Human Geography and GIS in the New Order

Human geographers have been pivotal in bringing ideas from STS and social theory into GIS over the past two decades. Though relations have been strained between these two subdisciplines at times in the past, especially during the 1990s when debates over GIS were on-going, there is increasing cooperation between the two. It is both an opportunity and convenience of sharing a discipline that ideas and theoretical constructs flow between subject areas as diverse as GIS and human geography. There is evidence of the productivity of this relationship in GIS with the development of three new areas: Critical GIS, feminism and GIS, and PPGIS.

Critical GIS is a broad rubric for an amalgamation of social theory and GIScience. It is characterized by its emphasis on extending the functionality as well as democratization of the technology. This dual social and technical emphasis has given rise to an agenda that is concerned with socioeconomic, participatory, epistemological, and algorithmic elements of GIS. Unlike human geography critiques of GIS, Critical GIS is an internal "movement" that includes scholars who have research interests that fall within the realm of traditional GIScience. In many cases, these scholars are cognizant of the role that social pressures have played in their own research, and want to acknowledge this aspect of the science they participate in. Francis Harvey and Nicholas Chrisman (1998), for instance, have documented the multiple ways in which a wetland can be defined for GIS analysis depending on the sociopolitical agenda of the actor. Environmentalists have defined great swaths of the continental USA as wetlands while prodevelopment and other factions frequently have much more limited visions of what constitutes a wetland. The definition and geographical demarcation matters intensely because wetlands are environmentally protected. Harvey and Chrisman propose that wetlands be envisaged, not as absolute areas that can be clearly outlined using crisp boundaries, but as *boundary objects* that are perceived and implemented differently depending on context and agenda. This allowed them to view wetland data as examples of interested visions rather than absolute reality – as many areas defined in GIS are regarded.

A number of GIScientists have been struggling precisely with the problem of how to represent multiple ontologies in GIS from a database perspective (see above). Researchers predictably approach the problem from different angles. David Mark and Barry Smith (1998; 2001), for example, believe that spatial entities that are known to exist should determine the ontological possibilities of GIS. They argue that nonexperts conceptualize geospatial phenomena using common primary spatial categories. The implication is that geographical terms hold common meanings that need to be accounted for when developing categories and spatial objects for GIS. For example, entities like mountains, roads, rivers, bridges, and towns are recognized almost universally as geographical objects that can be portrayed on a map. Such objects must be easily represented within GIS in order to allow nonexpert users to express spatial relations. These researchers recognize that there is an implicit "naturalization" of geographical categories in this empirically based ontology, but make the claim that there are fundamental, basic categories that are common sense.

Other GIScientists view language as more influential in defining ontologies. In their view, the fundamental issue is how to translate between feature attributes. Still others think that the issue of ontologies has been oversimplified. Andrew Frank (2001), for instance, suggests five levels of ontologies: human-independent reality (this level is consistent with Smith and Mark (2001); observations of the physical world that are obtained using measurement systems; objects with properties that are discerned through experience and cognition; social reality based on conventional names for things; and subjective knowledge. He argues that differentiating ontology types at the database level will permit the building of more consistent, reliable data as different parameters of accuracy and reliability are associated with different tiers of ontologies. Such a system accommodates natural categories as well as subjective knowledge. Francis Harvey (2003), however, argues that this schemata – based on realism – is only useful for constrained applications such as inventory management systems or air traffic control networks. Realist interpretations fail to take account of the vagaries of semantics, and the ways that language is interpreted differently in different environments. Each of these conceptualizations of language, however, constitutes part of a broader effort to improve the ability of GIS to represent natural and social phenomena – a fundamental aim of Critical GIS.

Feminism and GIS is a fledgling outgrowth of GIS with many of the same goals, but with a particular research emphasis that involves women as subjects as well as an examination of the extent to which women shape GIS at research and implementation stages. Mei-Po Kwan is a leading GIS researcher with many interests, feminism and GIS among them. Her earlier work examined the spatial patterns of women's movement outside the home based on travel diaries (1998; 1999). The space–time geographies of three different subpopulations of women were compared with respect to how far they traveled to work, and beyond the home–work route. Kwan's detailed analysis illustrated that women from all groups experienced more daytime fixity in their movements than men despite the fact that women employed full-time frequently traveled a longer distance to work than their male counterparts. An ameliorating factor was a greater number of household adults, presumably to share the chores. This research is one of a very few that specifically focus on women's space–time activities, and allows a rare glimpse into the gendered nature of spatial mobility. It also illustrates the value of using spatial analysis for areas of human geography that have historically been studied using qualitative techniques.

There is a growing complement of theoretical writing to buttress these efforts to use GIS for feminist research. Geraldine Pratt and I (2002) have written about the role of feminism in encouraging a more constructive dialogue between human geographers and GIS scholars. We argued that "how" critique is expressed, as well as what its objectives are, is critical to achieving changes in any research area. In the past, critics judged the processes and outcomes of GIS as problematic without grounding their criticism in the practices of the technology. We identified this as a pattern of external critique in which the investigator has little at stake in the outcome. Internal critiques, by contrast, have a stake in the future of the technology. Feminists such as Donna Haraway (1997) and Gayatri Spivak (1987) have argued instead for a form of critique that is invested in the subject, for critique to start from a position of caring about the subject. To be constructive, critique must care for the subject. Feminism and GIS is thus about engaging constructively with the technology as well as using it to do feminist research. Chapter 5 includes a broader discussion of the role of feminism and GIS in shaping possibilities of representation in GIS.

Feminism and GIS is also a more general call for engagement with technology. This call was taken up very early by the PPGIS movement. Daniel Weiner and Trevor Harris have taken an active part in PPGIS (Harris et al., 1995; Weiner et al., 1995). In one study, they investigated the role that GIS has played in recent changes in land distribution in South Africa. They were particularly interested in the extent to which GIS enables public participation and local knowledge. A GIS and Society project in Mpumalanga Province was the local focus of their investigation. Participant visions of local geography were mapped with colored pencils and then digitized into a community GIS. Questions included the spatial extent of forced removals during Apartheid, natural resource areas, ownership, land potential, and appropriate land use. The maps that resulted from these focus groups were striking, and illustrate the extent to which local knowledge can be integrated into GIS. They also provide the potential for making land reforms based on consensus (Harris et al, 1995; Weiner et al, 1995).

Other PPGIS researchers have asked the question: how can GIS be *rewired* to better accommodate public participation? Renée Sieber has written about ways of modifying GIS at a technical level to better accommodate the needs and working styles of nonprofit and grass roots groups. She argues that GIS would better serve these groups by accommodating overlapping and subjective spatial views often associated with local knowledges (Sieber, 2003). She suggest two approaches to achieve this end. The first is to encourage the use of UML (Unified Modeling Language) as a way for user to "map" or model the spatial objects that they are concerned with as well as a range of interactions between those objects. UML is notation or protocol for drawing objects and their relations in an object-oriented computing environment. Nonprofit groups could, for instance, *design* a GIS by defining features and behaviors of the spatial objects they define. Sieber also describes ways that XML (eXtensible Markup Language), a mark-up language for the web (like hmtl) can be used to include narrative descriptions of spatial features as well as pictures and drawings (Sieber, 2003). Such items are not easily accommodated by databases, but can still be integrated into GIS using XML. Sieber's approach illustrates the extent to which PPGIS research has not only brought GIS to community groups, but is also changing the technology.

All of these efforts represent the extent to which GIS use is dynamic and constantly changing, as well as the flexibility of the technology. Critical GIS, feminism and GIS, and PPGIS are answers to earlier concerns of critics of the technology. They are also illustrative of the extent to which the motivation of users determines the flexibility of epistemologies and ontologies. In effect, there are as many possibilities for representation in GIS as there are visions of geography and geographical relations.

3

The Devil is in the Data: Collection, Representation, and Standardization

In the previous chapter, an intellectual context for GIS was established – both from a disciplinary and implementation perspective. In this chapter the inner workings of GIS will start to become clearer as the data that populate GIS are described from multiple perspectives. For the uninitiated, GIS often appears to comprise multiple levels of analysis. It does, but analysis is only as persuasive as the data that underlie it. Moreover, data are not the transparent manifestation of reality in digital terms. They are the expression of particular points of view and agendas that begin as observations, and are transformed into numbers in data tables that provide the basis for spatial analysis. The first part of this chapter describes ways in which spatial data can be imbricated in political and social processes. How data are inscribed digitally is an important factor in its later use, and the conversion of textual and numeric information into tables is described in preparation for further discussion in Chapter 4 about spatial analysis. This process includes discussion of the metrics used to describe spatial entities including scale. The importance of metadata is also stressed as a vehicle for ensuring that data can be understood by subsequent users. In the past data were usually collected on a proprietary basis for a particular project. As GIS analysis extends its reach into multiple domains and scales of analysis, data are increasingly shared. In order to merge multiple sources of data, individual characteristics recorded in the data bases must be standardized. This process is integral to modern GIS, but remains fraught with difficulty. The logistics of standardizing data are described in the final part of the chapter. At the end of this section, newcomers to GIS will have a much clearer picture of how data are stored and standardized as well as their role in determining the outcome of GIS analysis.

GIS is clearly a social process and a complex combination of hardware and software. At the root of all analysis and representation are the data or the individual bits of information that, in combination, allow us to see patterns, build pictures of landscapes, and ultimately visualize events on the earth's surface – and underneath. These data are subject to social influences, as well as to the parameters of hardware and software. Data are collected by people using measurement devices or asking questions. They are then transferred to a computer, often by a different person than the original collector. Once rendered digital, they are subject to error correction, classification, standardization, aggregation, interpolation, and a myriad other possibilities for analysis. After they are manipulated, data are then engineered for visual output to the computer screen. One hundred or more things may happen to data on the way to becoming a GIS map. Indeed, if the initial data are of poor quality, it doesn't matter how much you manipulate them, they will still be subject to error. There are many techniques for compensating for data variability and error, but data quality remains the single best indicator of reliability for GIS results. "Good data" is an elusive category. It is dependent on many people and processes, and remains a fragile asset. It is more useful to think of data as an artifact that reflects people, policy, and agendas.

The Politics and Practicalities of Data Collection

The process of obtaining data starts with collection. Census data, for example, is collected in Canada through a form that is mailed to every known household. Recipients are required by law to answer the questions and return the form by mail. There are two forms that vary in length and detail of questions, and a statistical formula is used to determine how many long forms are sent out. The assumption is that information provided from the long forms will be statistically applicable to other homes in Canada. This premise means that some of the data reported in the Canadian Census are the result of statistical generalization rather than direct questioning. There is a high level of statistical reliability associated with this process, and, generally, Canadians accept the Census results.

Data are political, however, and this system of statistical extrapolation was the cause of feuding among US Democrats and Republicans in the late 1990s. A number of Democrat party members argued that the "count every man, women and child" approach used in the United States misses a significant segment of the population. Homeless people, the disenfranchised, and people who are suspicious of government intervention are among those who may not be counted. Census data are the basis for allocation of federal and state funds, as well as the demarcation of voting districts. A significant undercounting in the Census may mean that districts with the most need are underfunded and underrepresented politically. Those most likely to fail to return forms are also more likely to be Democratic supporters.

Democrats argued that a detailed Census based on statistical sampling will give the most accurate representation of numbers and composition of the American public. Republicans, however, gain more support from the middle and upper class. They countered that the US constitution stipulates that apportionment of tax dollars shall be dependent on population as reported by the Census, and *"The actual Enumeration shall be made within three Years after the first Meeting of the Congress of the United States, and within every subsequent Term of ten Years, in such Manner as they shall by Law direct"* (from Article 1, Section 2). Republicans considered this wording to be unequivocally in favor of direct enumeration. The Census strategy was not changed in 2000 despite charges that the Census continues to underrepresent significant numbers of people.

In Pima County, Arizona, near the Mexican border, officials estimated that 15,000 people were not counted in the 1990 Census. In that particular county, each person who was undercounted represented the loss of $2,000 to state and county coffers. This loss of $30,000,000 affected social services, schools, health care for the uninsured, road repair, and many infrastructure projects. The problem was considered sufficiently grave to warrant the formation of a grass roots organization to promote Census participation. The Pima County Complete Count Committee (CCC) worked to raise awareness among the undercounted population groups within Pima County, many of them recent immigrants from Mexico as well as Native Americans. They crafted a slogan "Kids Count, Don't Leave Them Out" to encourage people to report their children. Efforts were made to ensure that visible minorities were employed as Census field officers, and Census assistant officers were hired by the county. TV and radio commercials were played to encourage participation and numerous outreach projects were undertaken. "Demography is our destiny" said one grassroots organizer (Pima County Association of Governments, 2002). GIS researchers might counter that "data is our destiny."

The US Census recognizes the political stakes, and has taken steps to encourage traditionally underrepresented groups to contribute to the address database that the Census is based upon. The US Census Bureau uses a digital cartographic data bases called T.I.G.E.R. (Topologically Integrated Geographic Encoding and Referencing System) to store and describe address and demographic data. They developed a program called T.I.G.E.R. Improvement Program (T.I.P.) prior to the 2000 Census in which Tribal authorities and local governments were invited to update address files (Office of Research and Statistics, 2002). Existing maps for the

T.I.G.E.R. database were sent to officials who were asked to update roads, houses, and other features. This program illustrates the problems inherent in increasing representation: those most likely to be missed are not on the map. In March of 2001, US Commerce Secretary Donald L. Evans announced that the standard census data would be used to assign Congressional seats and distribute federal money, despite the fact that the Census Bureau's own estimates indicated that 3.3 million people were likely missed in the 2000 count. The numbers or data in a census, like the collection categories, are political.

The census represents one form of data collection – one that is used predominately in the social sciences. Other types of data are collected using numerous measurement tools and techniques. Precipitation data, for example, is collected using rain gauges. Sea-level depths are obtained through sonar imaging. Subsurface data are obtained from borehole samples and piezometers. Traditionally, human and physical geographers have relied on existing base coordinate data to describe the locations they are interested in. For instance, census tract definitions or postal code boundaries are frequently used as the basis of analysis. Even though the spatial definitions of postal codes and the census are the result of administrative fiat, and often don't match exactly the definitions of study areas, they are readily available in GIS format. Likewise, physical geographers frequently collect field data with reference to known ground control points such as weather stations for which latitude and longitude are available. Geographers, despite their spatial interests, have seldom had the survey tools available to determine unique spatial definitions for individual studies. Increasingly however, the use of global positioning systems (GPS) has enabled geographers to conduct primary data collection.

GPS is based on a system of twenty-four NAVSTAR satellites that circle the globe continuously. Originally developed by the United States military, GPS signals are available to anyone with a receiver. The accessibility of receivers has increased dramatically from those in the early 1980s that weighed over 50 pounds. Today, you can buy a wrist watch with a GPS receiver. Mountaineers routinely carry them as do police officers, gas line inspectors, field geologists, and a range of researchers. GPS technology is based on simple triangulation as illustrated in Figure 3.1. A signal is received from an orbiting satellite. The receiver measures how long it took to reach the earth. One signal permits a circle to be drawn around possible locations. Each signal provides more information, and allows the receiver to *zoom in* on a geographical location. Three is the minimum number of separate satellite signals required to provide a point location, and more are required for greater accuracy as well as elevation measurements.

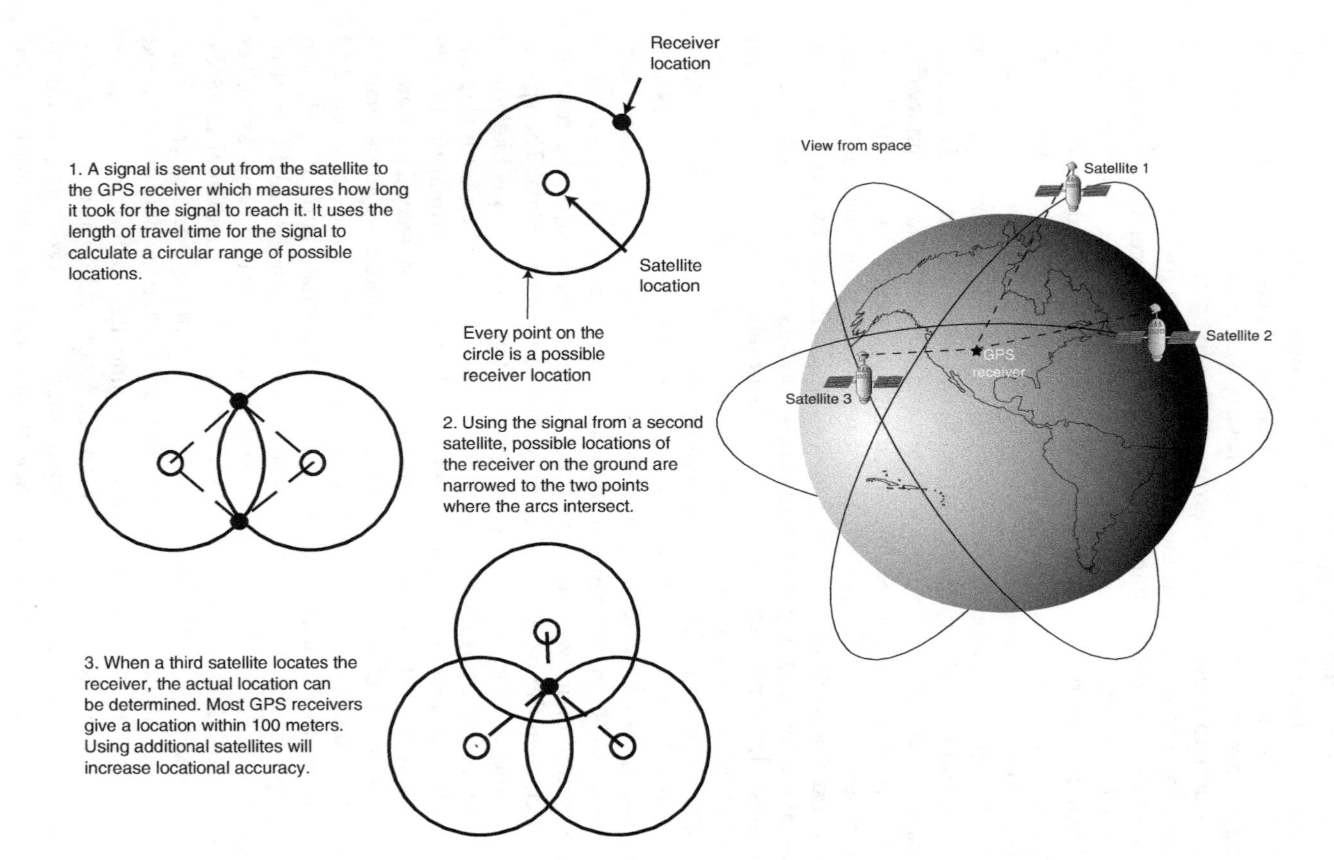

Figure 3.1 *How does GPS work?*

The range of accuracy has greatly increased since early GPS. In May 2000, the US military stopped distorting the signal leading to an increase in accuracy from 30 meters to within 10 centimeters and better. Differential GPS is the most accurate form of GPS, compensating for human and radio error by using two GPS: one roving, and another, stationary reference unit to monitor timing errors. If the fixed receiver has a known position, deviations from the measurement of the roving receiver (other than those based on distance from the differential receiver) are compensated for. In many cities, government agencies broadcast a differential signal that can be used to correct GPS signals. In Vancouver, as in many coastal cities, the Coast Guard operates this service. GPS is a democratizing force for spatial data.

GPS has contributed to the accessibility of spatial information, and, by implication, the ability of researchers to collect primary data. There is an important distinction between primary and secondary data. Primary data are captured using direct measurement specifically for use in the researchers' project. Secondary data are collected for another purpose, and may need to be converted for use in a GIS. Census data, scanned aerial photos, and digitized maps are secondary data sources. If you are interested, for instance, in the success rate of refugee resettlement in your area, you might conduct a poll among resettled refugees. Questions might touch on their current economic situation, their previous education, their level of English language training prior to immigration as well as their health. Clearly geographers would also inquire about their place of residence so that this information could be used for spatial analysis. If the questionnaire were designed and executed for a particular research project, then the results would be considered primary data. The spatial data used to describe the city or area of interest in a GIS is likely to be secondary data, but if GPS was used to locate refugee homes during the interview process, those point data would constitute primary data. GIS projects today often consist of a mix of primary and secondary data. In the early days of GIS, data capture often accounted for more than 85 percent of the cost of a GIS project. Today, it is between 15 percent and 50 percent thanks to increased availability of base data and GPS. Human geographers still rely heavily, however, on secondary spatial and attribute data collected through the census, socioeconomic surveys, as well as institutional data such as those from hospitals or environmental agencies.

We need to be particularly attuned to the challenges and discrepancies associated with these types of data. Because secondary data are collected for other purposes, the assumptions behind the creation of categories such as "community" or even "road" can vary widely as can the degree of spatial accuracy. There are various methods of accounting for differences in the meaning and accuracy of data, and they are implemented through data organization. In order to understand how data are influenced by political, policy, marketing, or economic agenda, it is necessary to understand how they are organized in GIS. The following section outlines the technical bases for storing and characterizing spatial data in a GIS. It is a skeletal primer on data, but permits a continuation of the discussion of social influences on data and the institutions that share them.

Organizing Data

GIS data, like data for all information systems, are stored in tables. The tables are organized differently depending on whether one is using a field or object data model (see Chapter 2). They look very similar, however, and they both list attributes of spatial phenomena including location. You might, for example, have a table that lists Canadian provinces with the population of each and the average income among other attributes. Another table might list roads in one particular province with the year that each was constructed, whether it is 6-lane, 4-lane, 2-lane, or a logging road. Tables with socioeconomic data frequently result from national censuses. A typical data table is illustrated in Figure 3.2. Tables are the basis for organization of data, and are configured to conform to

Attributes of CANcsd

FID	Shape	PRCDCSD	PROV	PROV_COD	PRCD	NAME	TYPE	PRCMA	CMA_CA	PRPCMA	PCMA_
0	Polygon	6204025	NU	62	6204	GRISE FIORD	HAM	62000	000	62000	000
1	Polygon	6107063	NT	61	6107	INUVIK, UNORGANIZED	UNO	61000	000	61000	000
2	Polygon	6204022	NU	62	6204	RESOLUTE BAY	HAM	62000	000	62000	000
3	Polygon	6204018	NU	62	6204	ARCTIC BAY	HAM	62000	000	62000	000
4	Polygon	6204019	NU	62	6204	NANISIVIK	SET	62000	000	62000	000
5	Polygon	6204020	NU	62	6204	POND INLET	HAM	62000	000	62000	000
6	Polygon	6107041	NT	61	6107	SACHS HARBOUR	HAM	61000	000	61000	000
7	Polygon	6108095	NT	61	6108	HOLMAN	HAM	61000	000	61000	000
8	Polygon	6204015	NU	62	6204	CLYDE RIVER	HAM	62000	000	62000	000
9	Polygon	6208087	NU	62	6208	TALOYOAK	HAM	62000	000	62000	000
10	Polygon	6107036	NT	61	6107	TUKTOYAKTUK	HAM	61000	000	61000	000
11	Polygon	6204012	NU	62	6204	IGLOOLIK	HAM	62000	000	62000	000
12	Polygon	6107014	NT	61	6107	PAULATUK	SET	61000	000	61000	000
13	Polygon	6208073	NU	62	6208	CAMBRIDGE BAY	HAM	62000	000	62000	000
14	Polygon	6204011	NU	62	6204	HALL BEACH	HAM	62000	000	62000	000
15	Polygon	6208081	NU	62	6208	GJOA HAVEN	HAM	62000	000	62000	000
16	Polygon	6208047	NU	62	6208	PELLY BAY	HAM	62000	000	62000	000
17	Polygon	6107017	NT	61	6107	INUVIK	T	61000	000	61000	000
18	Polygon	6107025	NT	61	6107	AKLAVIK	HAM	61000	000	61000	000
19	Polygon	6208059	NU	62	6208	KUGLUKTUK	HAM	62000	000	62000	000
20	Polygon	6208068	NU	62	6208	BAY CHIMO	SET	62000	000	62000	000
21	Polygon	6204010	NU	62	6204	QIKIQTARJUAQ	HAM	62000	000	62000	000
22	Polygon	6001043	YT	60	6001	OLD CROW	SET	60000	000	60000	000
23	Polygon	6107010	NT	61	6107	TSIIGEHTCHIC	CC	61000	000	61000	000
24	Polygon	6107015	NT	61	6107	FORT MCPHERSON	HAM	61000	000	61000	000
25	Polygon	6107012	NT	61	6107	COLVILLE LAKE	SET	61000	000	61000	000
26	Polygon	6208098	NU	62	6208	KITIKMEOT, UNORGANIZED	UNO	62000	000	62000	000
27	Polygon	6208065	NU	62	6208	BATHURST INLET	SET	62000	000	62000	000

Record: 1 Show: All Selected Records (0 out of 5984 Selected.) Options

Figure 3.2 *A data table with information about spatial features in Canada's north.*

database management systems associated with either object or field data models. There are several key elements of a spatial data set that transcend data models. They include location, consistency, scale, and metadata.

Location

There are endless numbers of data tables that can be constructed. What is essential in GIS is that each table include a spatial location. Without location, data are useless for spatial analysis. Hospital data, for instance, might include the date, patient registration, clinical symptoms, diagnosis, admission category, and date of discharge. This data could not be used in a GIS because there is no way of locating patients. If the data did include the home address or postal code of each patient, then an analyst would be able to discern a great deal of information. She would be able to determine whether people were coming to that hospital from outside its official catchment or service area. They might do this because of its reputation or because the hospital has an emergency ward, or because their local hospital does not offer a service such as stroke remediation. The analyst would also be able to test whether certain diseases such as measles are clustered around schools or daycare centers. With appropriate Census data, she would be able to determine if some medical conditions or emergencies are linked to neighborhood and individual profiles. Gunshot wounds may be clustered, for instance, among young men from a poor area. None of this is possible, however, unless the data include location.

Attribute data

Spatial data locate the geographic objects that we are interested in. Attribute data describe them. Attributes are characteristics of spatial objects defined in a GIS. If a province or state is defined as a spatial object, then population, average income, climatic extremes, and percent forestation might be attributes. In raster data, attributes are associated with individual cells, so the population of a province would be the sum of the populations associated with cells that represent the province. Attributes are described in database tables as values in a column (attribute) for a given row (records of spatial objects). What makes GIS attribute data different from that stored in statistical tables is that they are always associated with a spatial location. One of the keys to useful attribute data is consistency of collection and reporting practices.

Consistency

One of the great benefits of GIS analysis is that it permits researchers to examine patterns using large amounts of data over great areas. Global climate change analysis, for example, is only possible because of the enormous processing power associated with computing. Likewise the ability to create data sets that spans a vast territory is critical to such analysis. GIS has thus extended the scientific questions that can be asked, questions that were previously beyond the ken of researchers. There is a proviso, however, and that is that data must be consistent. There are many levels to this condition of consistency. At the level of the individual data table such as the hospital data, the same information must be collected for each patient. It won't do to collect most of the data but have a field (row and column in the table) missing for some patients. Missing data undermines the analysis. A few missing fields can be dealt with (often the notation – 9999 is used to indicate missing data). Likewise, data are usually associated with a particular time or period. Spatial phenomena are, after all, space–time entities. Analysis of an outbreak of Hepatitis A, for instance, is only legitimate over a period of weeks. Analysis of clustering over a twenty-year period would not tell the researcher anything about transmission or extent. Consistency is the key to good data, and good data are the key to reliable analysis.

Scale

Maps are the result of GIS analysis, and they are produced at various scales. Scale refers literally to the representative fraction that indicates what distance a given measurement on the map corresponds to on the ground. A scale of 1:5,000, for instance, means that 1 cm on the map represents 5,000 cm (or 50 m) on the ground. A large representative fraction associated with a map means that it is referred to as large scale. 1:5,000 is considered a large scale while 1:2,000,000 is small scale. Medium-scale maps can be anything from 1:50,000 to 1:500,000 depending on the context. Scale was traditionally very important in cartography not only because it provided a frame of reference for map users, but because it indicated appropriate uses and measures of accuracy associated with the map information. Small-scale maps include different types and amounts of information than large-scale maps. The former might be used for general overviews and familiarizing oneself with territory while large-scale maps provide rich local detail but cannot accommodate large areas. Small-scale maps are used for long driving trips, across North America or Europe while large scale maps are used to plan local forays where the placement of individual streets, local schools, and shopping areas is critical. Likewise, map users expect a far greater degree of accuracy (both spatial and attribute) associated with large-scale maps than smaller scale ones.

Scale has occupied such an important place historically in our collective consciousness that it was with great surprise that the GIS community greeted the insight of Michael Goodchild and James Proctor (1997) that scale does not exist inside the computer. Computers don't break up scale neatly into even numbers, nor do they analyze or represent information differently depending on scale. Scale is invisible during analysis from a computational perspective; it is simply a metric used to describe the display size. Indeed, the limited relevance of scale in the digital world is the basis for challenges faced by researchers attempting to develop automated generalization (see Chapter 2). Even if scale does not exist as a material entity in the computer, however, map projections and coordinate systems rely on scale. Without measures of scale, it is difficult to compare maps; it is also difficult to integrate spatial data unless their scale is known. Humans, unlike computers, rely on scale. We use scale to determine use and accuracy. Moreover, cognitive research suggests that humans perceive large-scale spatial entities (like houses, cars, lamp posts) as objects, but perceive smaller scale areas as landscapes. The problem is that scale is frequently intuitive and taken for granted by humans, but is literally oblique to a GIS.

This disjuncture can be ameliorated for display in GIS by giving a threshold above or below which features do not display. The user may decide, for example, that at a scale above 1:150,000, local roadways and residential streets should not display as they would clutter the screen, and be indistinguishable. Such techniques do not, however, circumvent the problem of scale with respect to data. Data are collected for use in certain applications at a particular scale. The widespread (and desirable) sharing of spatial data has led, however, to data collected for use at one scale being used for multiple scales, often inappropriately. Or data intended to convey a phenomenon over a limited area are generalized over a large area. Data collected for individual enumeration areas, for instance, are frequently aggregated to Census tracts, in the process losing the nuances of population density or income distribution. This is a normal and intended scale shift, but illustrates the way that a change in scale can obscure information. Figure 3.3 illustrates how the spatial footprints change as enumeration areas are aggregated to census tracts, and then to census subdivisions. GIS users, however, also make inappropriate scale shifts with data. When population density data for an entire province are conveyed on a very small-scale map, the aggregation is

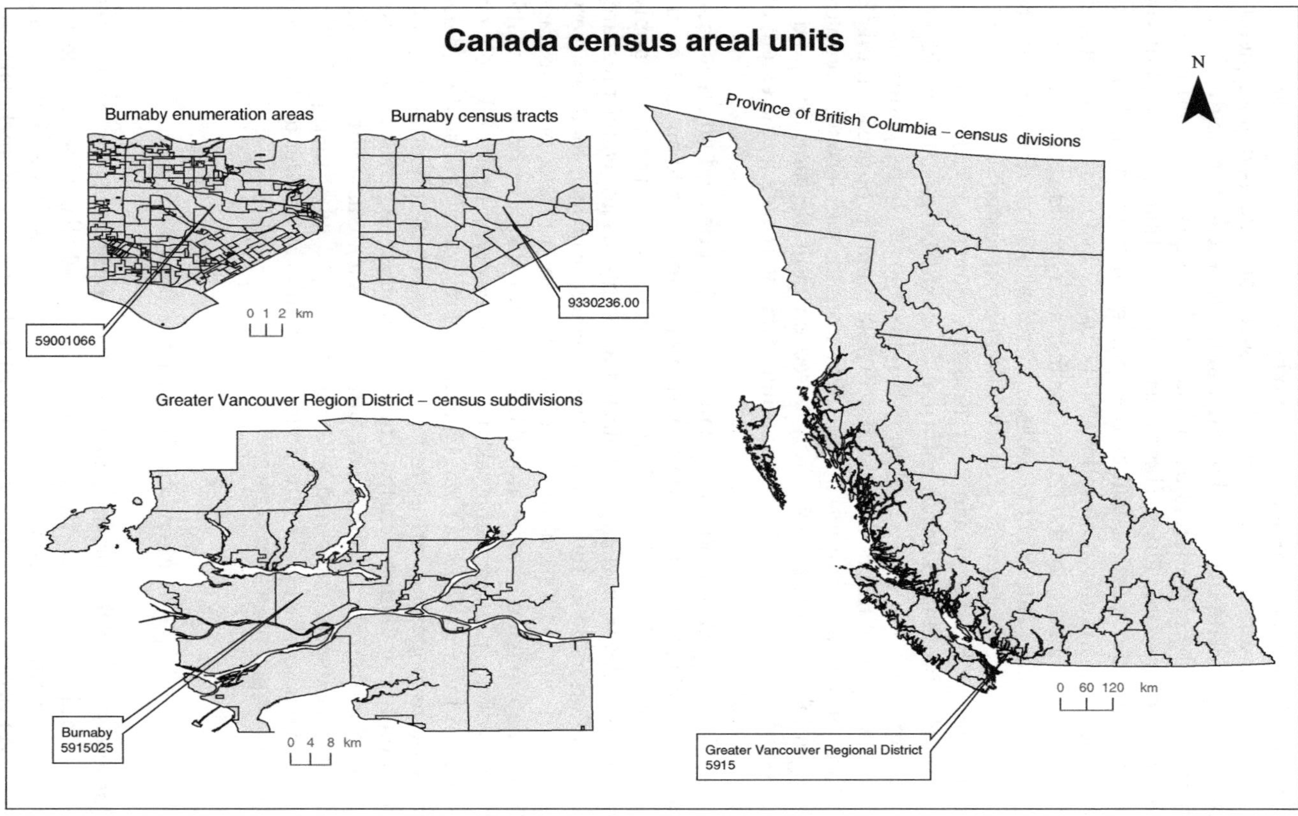

Figure 3.3 *Aggregation of spatial footprints often associated with census data. Source* Statistics Canada

justified, but when a large-scale map shows population density spread evenly over a town, this is misleading as population tends to cluster, and some neighborhoods are denser than others.

Point data are particularly subject to issues of scale as it is frequently used to interpolate between known values. A researcher might, for instance, know the elevation of 30 points within a square km, and want to discern intermediate values. There are many spatial statistical methods of interpolating the missing values depending on how the initial points were sampled and the assumptions of the user. Given this point density, a reasonable estimate of the terrain for that area would result. A fault or small mountain peak might be missed if critical high or low points were not sampled, but the general trend would be represented. If just 30 points were known for the province of British Columbia (BC), however, it is unlikely that any relevant information could be gleaned about the topography of the region. Figure 3.4 juxtaposes two renditions of the topography of BC. In the first, the 30 data points produce a simplistic picture of the topography of the province. In the second, complex changes in elevation are evident – based on multiple sampling points. For point data, scale relates to sampling density as well as the extent of the study area.

Despite the widespread use of data models that depend on areal data, most GIS data start as point data, and are extrapolated to areas. Census data are collected from individual households, and aggregated to apply to a region. Elevation data are collected at points, as are water pressure, noise pollution data, and rainfall measurements. In fact, there are very few data used in GIS that do not originate as point data, but our traditional map-making devices are all based on the display of areas. In GIS, these areas are polygons and rasters, and even object-based GIS rely on vectors or rasters to display spatial entities. Points are almost always the basis of homogenous areas representing levels of phenomena from elevation to population density. The question then becomes: how do we know where our data come from? Did they originate as point data, were they aggregated or combined with other data? To what map projection does the data correspond? Who collected them? The answer to these and other questions lies in metadata.

Metadata: Data About Data

Questions about the origins, quality, and applicability of data were largely irrelevant in the early days of GIS as almost all data were collected for specific purposes such as to make land use maps or designate zoning areas. As people and organizations started to share data, metadata – or

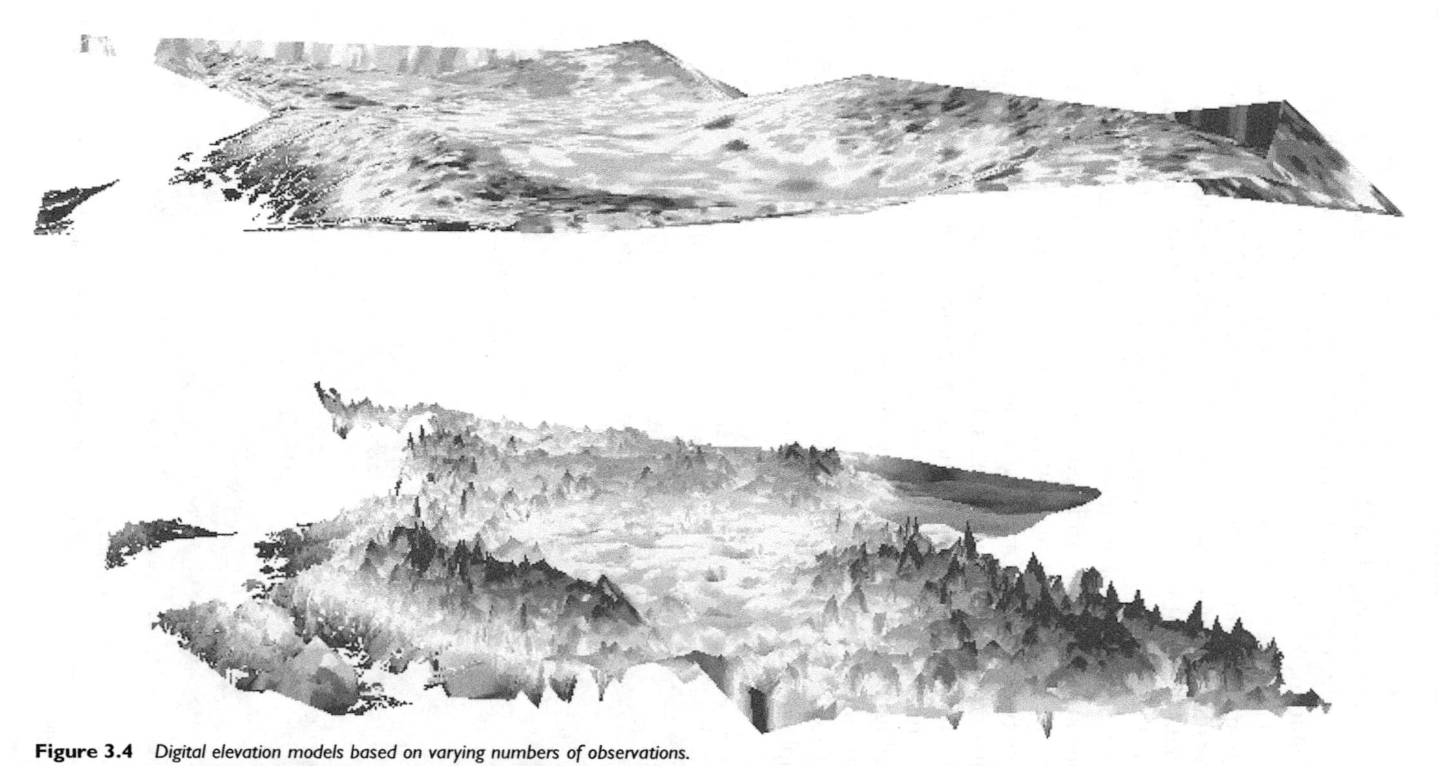

Figure 3.4 *Digital elevation models based on varying numbers of observations.*
The top map illustrates the topography of the Province of British Columbia based on only 30 elevation observations. The bottom map is based on an intricate web of data points and better reflects changes in elevation.

data about data – became increasingly important as a means to assess the compatibility among data. Metadata answers questions like: When was the data collected? What time period does it cover? What scale are the data applicable to? What projection was used? How were the data collected? What quality measures were taken? How were the data classified? And what mapping units are associated with the data? Metadata – or a list of answers to these questions – is usually included as a text file attached to map data. Without metadata, it is very difficult to combine datasets that were collected by different organizations or individuals. Data sharing is essential as a means of broadening analyses. It allows users to map distributions across borders and between jurisdictions. Metadata is essential, for example, to create maps that compare the number of people with university degrees in Canada and the US or to compare use of nitrates in farming across county boundaries. In both instances, metadata provides information about data scale, quality, and temporal relevance.

A number of initiatives, especially in the US, have led to metadata becoming an accepted – and expected – dimension of spatial data. GIS programs like Clark Labs IDRISI have an interface that supports transfer and supplementation of metadata as illustrated in Figure 3.5. This example of metadata provides primarily information about the spatial

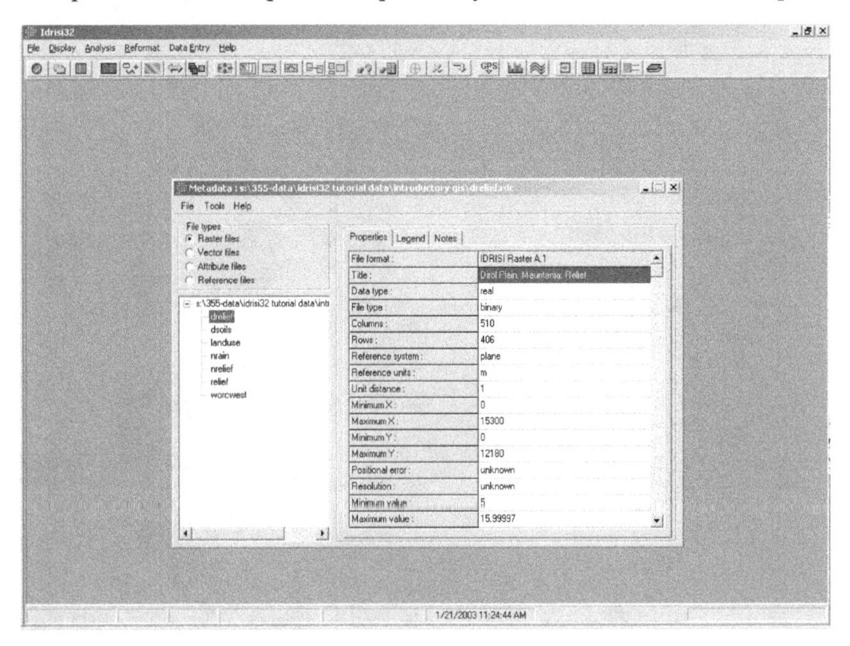

Figure 3.5 *Metadata from IDRISI.*
Metadata is data about the data used in GIS and contains information about the suitability of the data for a specific purpose.

components of the dataset. Recent research has focused on ways of embedding information about the ontologies and epistemologies associated with attributes, but these efforts are not yet reflected in off-the-shelf software programs. Support for metadata standards to describe spatial data is frequently offered by national organizations with the express mandate of developing spatial data infrastructures. In the US, the National Spatial Data Infrastructure (NSDI) is charged with developing spatial data formats; in Canada GeoConnections is the umbrella body responsible for creation of the Canadian Geospatial Data Infrastructure (CGDI). Such infrastructures far surpass metadata; they are being developed to allows seamless sharing of data across computing environments and jurisdictions.

Sharing Data: Interoperability

Interoperability is the pursuit of a common language for computational environments, based on a foundation of common spatial grammar. It is a rubric for negotiation between systems and information. Perceived benefits include reduced costs and time to transform and manage data. Even with metadata, it is a time-consuming and costly process to merge data from multiple sources. Data interoperability emerged as an ideal because clear principles for the consistency of GIS data and databases are the basis for sharing data worldwide, and allowing an increase in the scope of GIS analyses. Similar to the open architecture plan, introduced by IBM with the first PCs in the 1980s, and the current promise of Linux, interoperability harbors the aspiration that common specifications will seed and enhance the market. Interoperability will cultivate transitions between software and hardware environments. In principle, it will allow data to be used across platforms without any loss of integrity. It also promises to reduce the cost of data acquisition and management.

Interoperability addresses not only data compatibility but systems and network compatibility. The most difficult task, however, remains semantic interoperability. Semantic aspects of databases are those that have to do with language. Merging data is so difficult because even when data collectors use common terms, such as "urban" for built-up areas and "rural" for regions of low population density, their working definitions vary. David Mark a prominent GIS researchers at SUNY Buffalo illustrates the difficulty of semantic integration with the example of the word "pond." One might imagine that integration of two different databases describing ponds would be straight-forward, but Mark points out that even within the country of Canada, pond is used and interpreted very differently. In Ontario, for instance, pond is interpreted as a small lake, but in Newfoundland, a pond can refer to a very large body of water with characteristics that are associated with lakes in Ontario (Mark, 1993). Semantic heterogeneity persists even when geography and language are shared.

Consider a mapping project of the vegetation within a forested area. The project is carried out by two different teams of professionals. The first team consists of wildlife biologists and the second of foresters. Both teams have extensive local knowledge, education, and experience with collecting vegetation data in the area, yet their results will differ significantly. Wildlife biologists tend to use a classification system that highlights the importance of vegetation to wildlife (i.e., open-range area for grazing). Foresters, however, use a classification system that documents commercially viable tree species. The two classification systems might contain similar features (fir trees), yet the interpretation of the similar features could be quite different. The wildlife biologist is likely to be concerned with crown closure,[1] since it contributes to snow cover during the winter months, and important for wildlife survival. The forester, however, would be interested in the height and diameter of the fir trees, since this contributes to their commercial value. Not only is it expected that the different communities (wildlife biologist and forester) use semantically heterogeneous terms, but it is also conceivable that within each community there will be semantic differences. How can the semantics of feature abstraction be made homogenous within user communities and how can the semantics of one user community be evaluated against another?

The example of roads in two different databases developed by different agencies within the Canadian province of British Columbia (BC) is instructive. Several BC government ministries have separately collected and maintained environmental and forest resource data sets. Despite detailed common procedures for identifying features, there are considerable differences in their classification schemes. The storage of roads provides a concrete example. The BC Ministry of Forests (MOF) *Forest Cover* dataset contains more roads than that from the Ministry of Sustainable Resource Management (SRM) as illustrated in Figure 3.6. The road data do not match because each Ministry maps roads in different places, based on separate field work. Even when the roads are identically defined as spatial features – as illustrated in Figure 3.7 – the semantic definition of what constitutes a road can vary between institutions. The same roads can be defined differently even when they represent the same feature.

[1] Crown closure is the density of the forest canopy, usually represented with a numeric value. 100 percent indicates that there is no open space within the forest canopy.

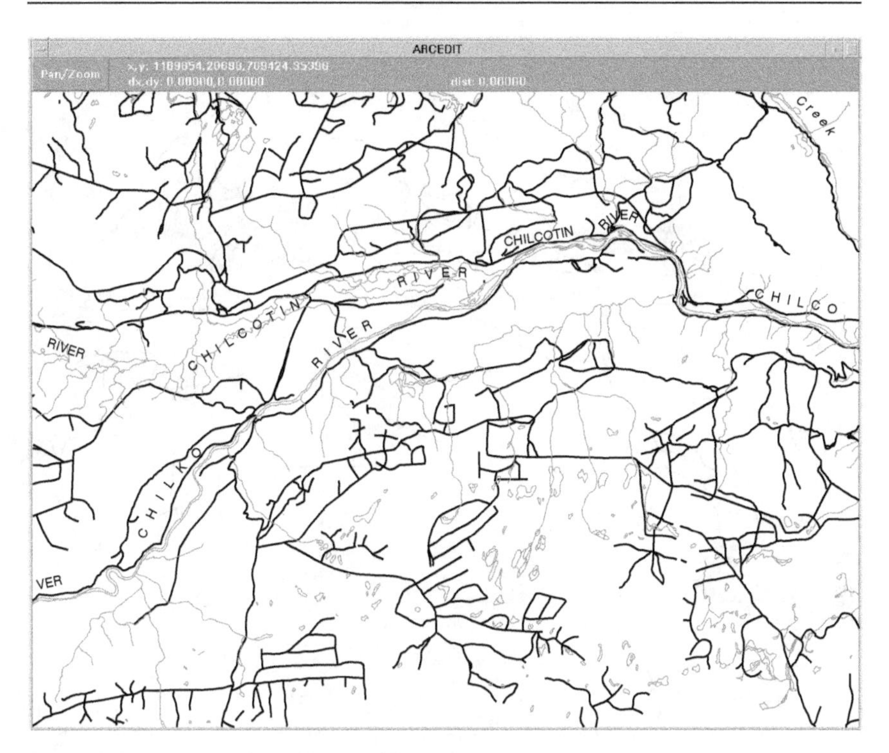

Figure 3.6a *Road data for the Ministry of Forests.*

Source: *First printed in* Cartography and Geographic Information Science, *vol. 29, issue 4.*

In this case, the definition of road varies between agencies. The Ministry of Forests definition uses the following definition (see: http://srmwww.gov.bc.ca/gis/imwgstd.doc):

DA25150000 – A specially prepared route on land for the movement of vehicles (other than railway vehicles) from place to place. These are typically resource roads, either industrial or recreational. Includes road access to log landings.

The Ministry of Sustainable Resource Management, on the other hand, uses this definition:

DD31700000 – A narrow path or route not wide enough for the passage of a four wheeled vehicle but suitable for hiking or cycling. Park paths and board walks are considered trails.

These differences, though ostensibly minor, can make it very difficult to share data between government agencies in the same province let alone across national boundaries or between a government ministry and a grass-roots environmental group. Such semantic discrepancies

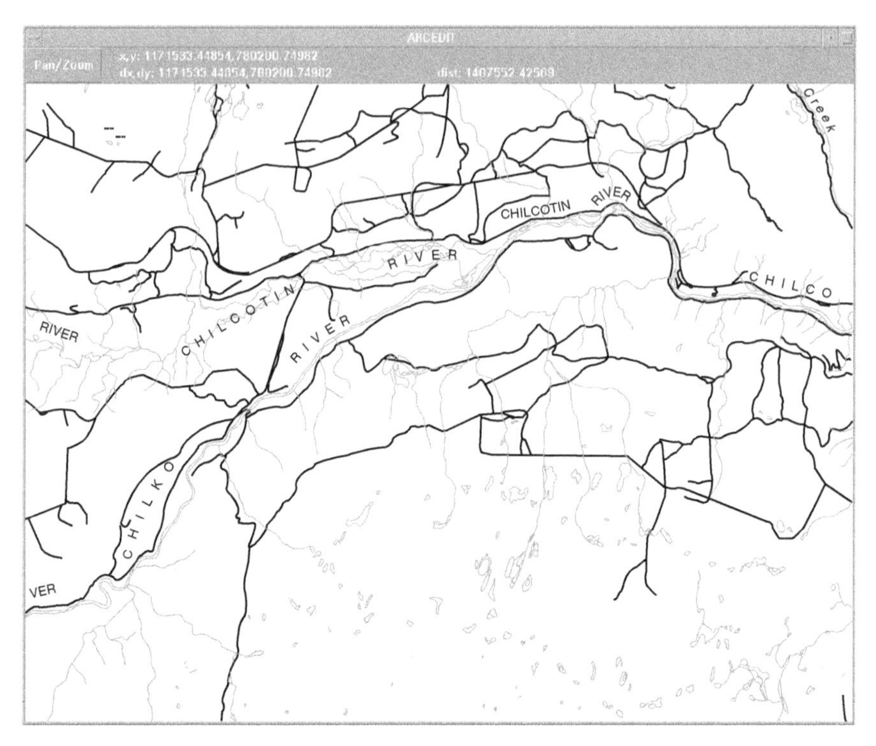

Figure 3.6b *Sustainable Resource Management road data for the same area as Figure 3.6a.*

Source: *First printed in* Cartography and Geographic Information Science, *vol. 29, issue 4.*

can undermine digital data sharing in the decentralized data collection environment found at many levels of government.

The problem of semantic heterogeneity is intensified as the complexity of the phenomena increases. In situations where multiple data sources and attributes are combined to make decisions about geographical extents, the problems of language and data are compounded. In the province of British Columbia for example, the Mule Deer requires a protected habitat for winter grazing. The Mule Deer is concentrated in the Cariboo Region of the province. This is a variably forested area with relatively high snowfall, and cold winter temperatures. In order to protect winter grazing ranges, information about tree species, crown closure, and the number of tall trees within a site is required. These three attributes are often collectively referred to as "stand structure" data. In this case, an extant database with stand structure information for some of the designated winter ranges was already in use. Ideally the Ministry of Environment, Lands and Parks (MELP) would have collected stand structure data for the remaining habitat zones, but time and money

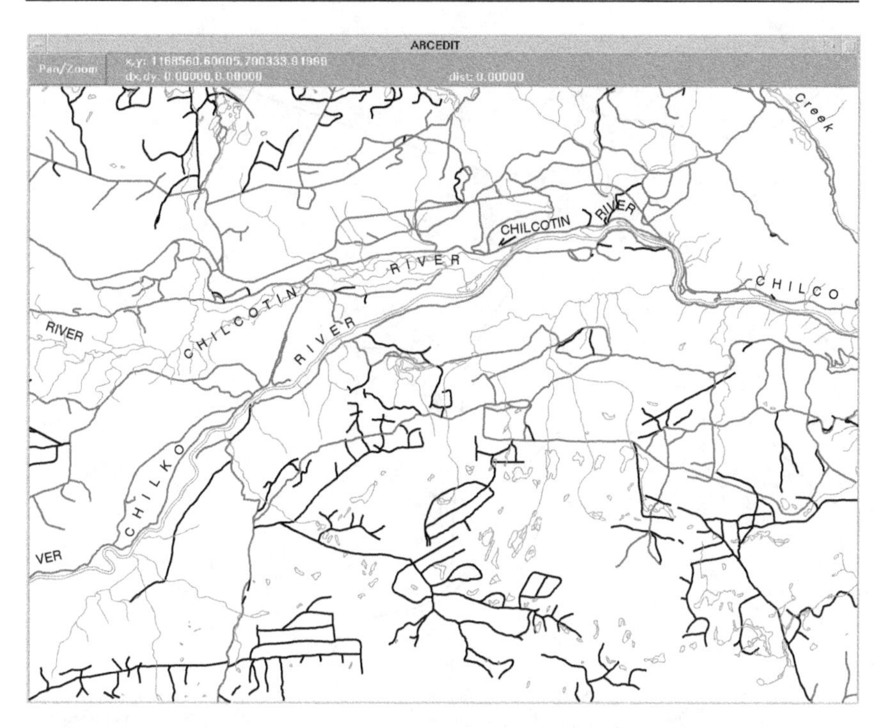

Figure 3.6c *The Ministry of Forests road data shown as thin lines and Sustainable Resource Management roads shown as thick lines. (See http://www.blackwellpublishing.com/schuurman for color version.)*

Source: *First printed in* Cartography and Geographic Information Science, *vol. 29, issue 4.*

restrictions prevailed. In order to complete a land use plan for the entire region, the agency decided to use existing forest cover data from the Ministry of Forests by integrating them into their existing database.

Each attribute from "forested land" to "snow cover" is defined and understood differently by different agencies and researchers. As a result, attaining agreement on the definition of geographical attributes in order to convey a complex concept such as habitat is a challenge. In this case, MELP initially used aerial photography at a scale of 1:15,000 to classify their original stand structure data. Using the imagery, they were able to define snow pack zones, slope, aspect, and crown closure in the forest overstory (the layer of foliage in a forest canopy), as well as the number of large trees (Douglas Firs) per hectare. The forest cover dataset maintained by the Ministry of Forests Resources Inventory Branch (MOF) was obtained using different methodologies. The reference data were originally mapped at a scale of 1:20,000 in 1972 while attribute data were collected at different times – the product of various field projects.

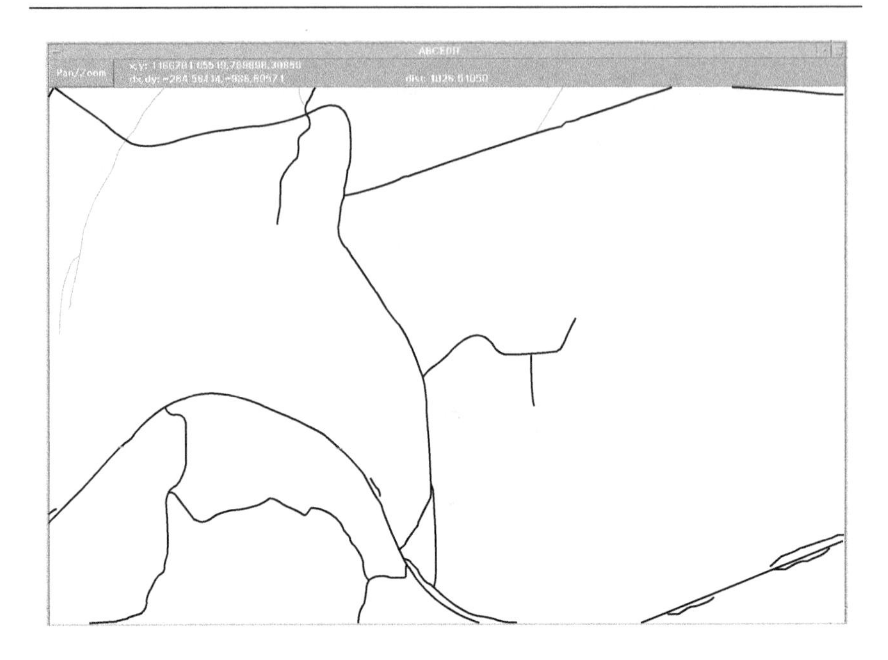

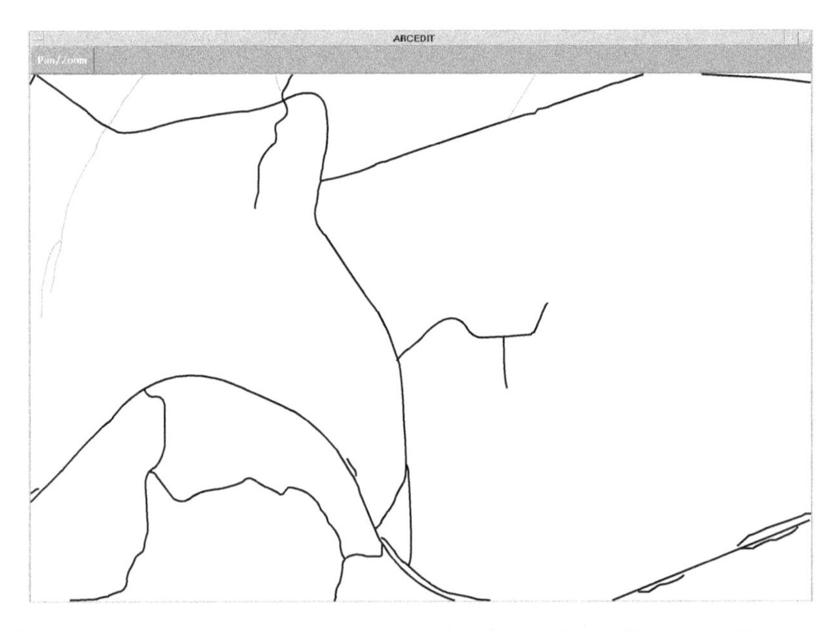

Figure 3.7 *The Ministry of Forests road data juxtaposed to the Sustainable Resource Management road data for a portion of British Columbia. Even though the two road coverages unmistakenly represent the same features, and the cartographic representation is identical, the roads are defined by the two Ministries in semantically different ways.*

Dissimilar data collection methods contribute as much to semantic heterogeneity as do differences in the usage of language. Each data collection project is developed with different goals, and is associated with different epistemologies. Biologists, foresters, and GIS specialists each report the world and its attributes differently. Given the exigencies of time and money, however, the decision to proceed with the data integration was made. Further problems ensued. The Ministry of Environment collected their data with the express purpose of identifying viable areas for the Mule Deer to graze. The forest cover data from the Ministry of Forests was collected primarily for forest management purposes. These divergent goals resulted in different definitions of stand structure attributes. The definition of "crown closure," for example differed between Ministries. The MELP definition was based on the amount of ground area covered by the tree crown. Only coniferous trees with a pole height greater than 10.4 m were included. The MOF definition did not exclude deciduous nor short trees. Moreover, their data for this area dates from 1972. The data were thus semantically and temporally heterogeneous.

Not only were the data discordant, but the two Ministries used different classification systems. Table 3.1 illustrates how the attribute of crown closure varied.

The effects of these different classification schemes are illustrated in Figures 3.8a and 3.8b. Each map provides a very different account of the extent of the forest that has full crown closure. The multiple problems associated with this data integration project demonstrate the axiom that data are never naïve. They are reflections of beliefs and processes. They are frequently social, and always reflect a particular epistemology. Even if they correspond semantically, they may be differently classified, and categories themselves are expressions of epistemology. Semantic

Table 3.1 *Crown closure classification*

Class values	*Forest cover crown closure values*	*Stand structure crown closure values*
0	0–5 %	n/a
1	6–15 %	0 %
2	16–25 %	1–15 %
3	26–35 %	16–35 %
4	36–45 %	36–55 %
5	46–55 %	> 55 %
6	56–65 %	n/a
7	66–75 %	n/a
8	76–85 %	n/a
9	86–95 %	n/a
10	96–100 %	n/a

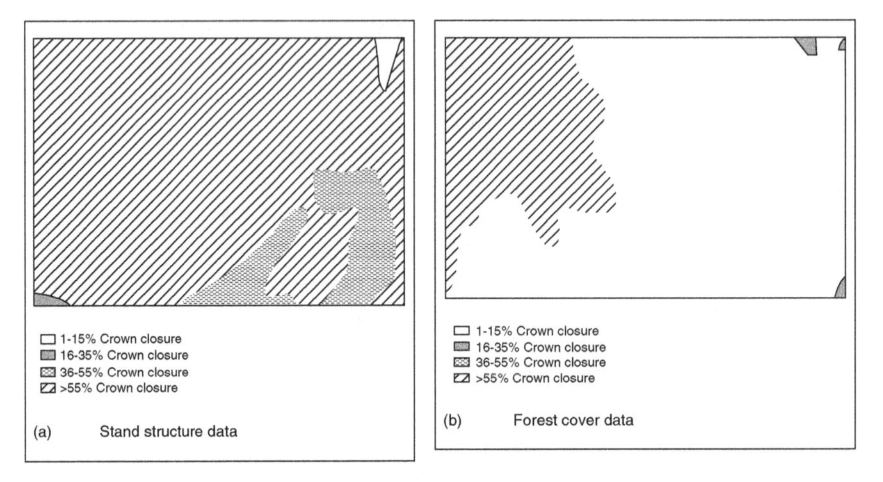

Figure 3.8 *(a) Stand structure crown closure values. (b) Forest cover crown closure values.*

differences between domains or areas of interest make interoperability a continuing challenge in GIS.

Despite the immense problems of reconciling the meaning of disparate spatial data, it is inevitable that data be integrated from multiple sources. It is not economically feasible for every GIS project to acquire proprietary data for specific projects. Standardization is the process through which disparate terms for similar – but rarely identical – entities or attributes are homogenized in order to use multiple sources of data. For example, two different data sets that describe soils might use terms such as permeability or loam very differently. If the initial descriptions of criteria used to define the data are available from the metadata, then it is possible to reclassify data from multiple sources to more closely match the meaning of attributes. For example, the definition of road used in the Forest Cover data set (see above) is very different than that used by the Sustainable Resource Management group. Despite the fact that the roads frequently refer to the same feature, their semantic and physical descriptions vary depending on context and interpretation. There is a clear disjuncture between categories (definitions) and instances (defined roads). This is likely to be the result of both broad interpretation of road definitions by operators, and different institutional cultures between the Ministries of Forests and Sustainable Resource Management. In order to standardize the term road, it is necessary to not only have access to the definition used for road, but understand how "road" was interpreted in different institutional settings.

Likewise, when joining data tables from different sources, there is no way to identify fields that refer to the same entity by different names. "Park" and "recreational area" or "logging road" and "limited access road" might be equivalent, but would remain separate categories unless flagged by an operator with extensive local knowledge. Unfortunately, few GIS users are aware of the possible complications of combining data, nor do they have the time to conduct institutional ethnographies before merging datasets. The result is less reliable data and, by implication, less reliable analysis.

The importance of semantic standardization is, however, recognized by the GIS academic community, and many scholars are developing methods of standardizing data from multiple sources. A number of approaches have been developed including federated data sharing in which an automated mediator matches terms from multiple data tables. Another approach uses concepts from linguistics and computer science to automatically identify terms that share similar descriptions. A variation of this approach is to look for terms with similar contexts and create equivalences between them. Each of these methodologies is, however, limited by their emphasis on cutting edge database technology (based on object orientation), and the assumption that complex relationships like "context" or "equivalence" can be understood and implemented by a computer. Routine users of GIS are left with the problem of joining tables from multiple datasets in ways that maximize the relevance of the data for a particular project. The challenges of data collection, standardization, and classification continue to challenge decision makers. The variability and vagaries of data affect the viability of much GIS analysis as the example below illustrates.

Collection, Standardization, Classification: The Example of Groundwater Data

The complexity of data handling, standardization, and classification is conveyed by the example of well-log data in Canada. Environmental specialties are proliferating in human geography departments as threats to the environment threaten our standard of living and quality of life. The relevance of well logs to human geographers may, at first, seem elusive, but presents an example of the relationship between data, analysis and study areas. Human geographers are frequently interested in ascertaining areas of environmental vulnerability. Water well logs are a window into the structures that form the subsurface of the earth; they constitute the data used to determine where aquifers and aquitards (impediments to the flow of water) occur. Figure 3.9 illustrates how well logs are used to create schematics to represent what lies below the earth's surface. Earth scientists can use the information from multiple cross-sections to develop models of the direction of water flows, and where aquifers are likely to exist. Knowledge of where an aquifer is and how it is fed constitute vital information for assessing environmental vulnerability. Storage areas for groundwater are normally considered protected zones because they supply water for drinking, agriculture, fish streams, and recreation areas. A partial list of industries and activities that rely on information about aquifers and aquitards includes drinking water administration, city planning, waste burial, agriculture, real estate development, and environmental resource management.

Water, like other resources, is considered the property of the provinces in Canada, and is managed at the provincial level. Groundwater management in most instances is based on well-log data that is collected by individual water well drillers across the country, many of them privately employed by home-owners or businesses. The level of expertise associ-

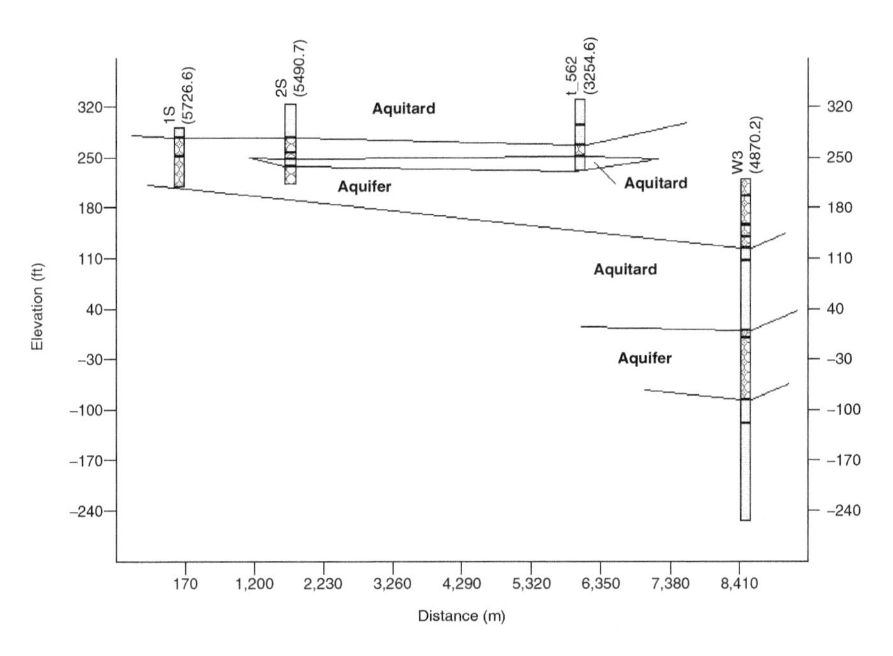

Figure 3.9 *Cross-section based on well-log data.*
Whenever a borehole is drilled, a well log is recorded with some lithological information. These data provide the basis for an understanding of what lies beneath the earth's surface. In this case, materials that permit the flow of water have been classed as "aquifer" and bedrock and impermeable materials as "aquitard." These designations are linked together to create simple reconstructions of the subsurface lithology.

ated with the data collection varies widely. In Newfoundland, there are only seven licensed water well drillers, and each is trained and certified by the provincial Ministry of Environment and Labour. By contrast, drillers in British Columbia are not required to complete any training before identifying subsurface layers, and reporting them to the provincial government. Differences among provincial Water Acts also contributes to the variability of the data. In some provinces, drillers are required by law to submit any well logs to the provincial government. In others, submission of well logs is done on a voluntary basis. Likewise, the data are available for academic research on a province by province basis, and availability is also dependent on whether the well logs are in digital form. Nor is this problem confined to Canada. Many part of the US rely on well-log data from private drillers to manage subsurface water. Despite the inconsistency of the data reporting, they are of vital importance for environmental resource management.

GIS modeling of the subsurface begins with well-log data which is gathered in the field by drillers. Drilling for water is harder than it looks, and identifying the various geological strata or lithologies as the drill is moving is even harder. Mud is used to lubricate the drill, and as a result much of the subsurface material looks like mud. A study in Ontario found that over 40 percent of the subsurface material was identified by private drillers as "clay." Control drilling by scientists from the Geological Survey of Canada (GSC) identified clay as 2 percent of the material in the same area. Indeed, inconsistency in the identification of subsurface lithologies is an even greater problem than availability of well logs. Few drillers are trained to identify geological strata, and the scenario of drillers relaxing in a doughnut shop filling in the forms after drilling is not inconceivable. The range of education and training of drillers has resulted in extreme *semantic heterogeneity* in the data.

The extent of these semantic differences is illustrated by British Columbia water-well data. In this province which admittedly encompasses a range of geological areas, there are over 53 different material terms used. These are prefixed by 41 different adjectives. The rules for combinations and permutations suggest the variety of possible lithological descriptions that can result. Indeed, as Table 3.2 below illustrates, it is common to see thousands of different terms used in a small area of the province.

Table 3.2 identifies two semimountainous regions (the Cariboo and the Kootenays) and one small island. In each case the number of wells is surpassed by the number of lithological units because several samples are taken at different depths for each well. The table illustrates the extent to which the descriptions of similar lithologies vary depending on the drillers and their interest and experience. In the case of the Cariboo region, for example, there are over 5,000 different descriptions. Cartographic

Table 3.2 *Degree of data heterogeneity for three geologically distinct regions of British Columbia*

Region	*Number of wells*	*Lithological units*	*Unique lithological descriptors*
Cariboo	2,726	15,113	5,195
Kootenays	2,350	8,717	3,860
Galiano Island	859	3,773	1,179

research has shown that map readers cannot understand more than seven different categories on a map. Certainly 5,000 would challenge even the more sophisticated user. So the first problem is how to reduce the number of categories while retaining the inference intended by the well driller. The second challenge is to standardize these data across the province, the country, or even the continent.

Standardization is very closely related to classification because in order to standardize data, one must create categories or classes of spatial entities or attributes. In this case, the subsurface is an enormous information space that must be described by a handful of terms. The choice of those terms constitutes the process of classification. We tend to think about classification systems as being absolute and certainly accurate, but they are developed by individuals or institutions that see the world in particular ways. Just as foresters and wildlife biologists classify old-growth forest differently, a hydrologist will classify lithologies very differently than a petroleum geologist. These differences arise because each domain is interested in different information about the phenomena. Diverse ways of seeing the world are, however, only part of the problem associated with standardization and, by implication, the categories of analysis for GIS.

Classification, like data, is a political process that reflects the exigencies and cultures of particular institutions. GIS uses well-defined spatial entities especially when in a vector or object environment. These entities are described using strict, linear boundaries, and give the impression of having crisp delineation. In actual fact, the boundaries of such objects from forests to urban areas are often controversial. It is the process of classification that has determined their extent. The internal structure of a classification system reflects agendas that are ultimately discerned through studies of practice. Geoffrey Bowker and Leigh Star are researchers in Science and Technology Studies (STS). They have conducted detailed studies of the ways that classification systems emerge. Bowker and Star (2000) have illustrated, for instance, that the International Classification of Disease (ICD), used to categorize medical ailments, privileges the perspective of doctors over patients and legal discourse over social custom and patient perspective. Bowker has also documented ways that biological classification systems reflect beliefs of a period, rather than absolute truth in science. Despite the limitations of classification, it remains a necessary step for standardizing data to use in GIS. It also provides a basis for creating objects for analysis.

In GIS, as in other types of information science, there is a tendency toward pragmatic, situation-based classification rather than broader, more comprehensive models. Users of a system, for instance, may develop the number of categories to correspond to the maximum legend length in a particular software application. Bowker and Star illustrate this tendency with the example of the ICD which had an original limit of 200 categories – that corresponded exactly to the length of the Austrian census forms. This pragmatism is not in itself wrong, but should be recognized particularly because there tends to be a naturalization of classification systems. Over time classification systems begin to be perceived as true and inevitable. The politics of classification are forgotten once the system is in place. The challenge with water-well data is to classify in order to render the data viable for environmental analysis while retaining user perspectives on the "situatedness" of classification.

Standardization of water-well data involves making decisions about what categories to use, and how to divide existing attributes into those categories. There have been several attempts to standardize classification systems for the entire country of Canada. These have not been successful for political as well as scientific reasons. Water is considered a provincial resource, and is managed by 10 governments. Agreement on classification entails bringing multiple partners with multiple levels of interest to the table. Discussions among the 10 Canadian provinces about a groundwater classification in 1991 broke down over the issue of whether to include the word "national" in the title. In this case sovereignty issues rather than nuances of hydrogeology created the impasse. The most successful attempts to classify well-log data have been achieved in Canada at the regional and provincial level. The Oak Ridges Moraine (ORM) Project developed in concert with the Geological Survey of Canada (GSC) and the Ontario Ministry of the Environment (MOE) offers a successful example of using classification to enable environmental modeling and planning based on well-log data.

Water well-log data were the basis for an in-depth study of aquifer vulnerability in the Oak Ridges Moraine, that includes the Greater Toronto Area (GTA) in Ontario. The ORM project was designed to give earth scientists, city planners, and environmental managers a blue-print of the subsurface of Canada's largest metropolitan area, an area that houses over seven million people as well as many industries and intensive agriculture. The project began by standardizing well-log data based on a classification scheme developed by in-house experts at the Geological Survey of Canada (GSC). These data were used to create three-dimensional models of the subsurface in which aquifers and aquitards were identified. Given the variable quality of the well-log data, control studies were conducted to ascertain the reliability of data, and the results were incorporated into the models. The models were subsequently used for city planning, waste management strategies, and modeling research in government agencies. They were also artifacts in a later inquiry into water contamination in the town of Walkerton, Ontario (see below).

The ORM standardization scheme is based on the creation of 12 categories of subsurface material. Each of the 12 categories is further subdivided so that the standardization is effectively associated with different granularities or resolutions of data as illustrated in Figure 3.10. Users who require the very detailed descriptions can subdivide the basic categories. Given the variable quality of the original well-log data, the broader description is usually more appropriate. In cases where control data are available greater differentiation of detail may be desirable. These data were used by scientists at the GSC to create conceptual models of the geology of the ORM area as well as regional three-dimensional models of the hydrogeology with emphasis on direction and flow of groundwater. Their work is the basis for environmental management of water resources in the rapidly growing Greater Toronto Area. Ironically as the GSC scientists accomplished the massive task of making these well-log data usable for modeling and environmental inquiry, they unearthed new logistical and policy-oriented problems.

The ORM data are part of large volumes of data being collected by federal and provincial governments in Canada and internationally. In most countries, government agencies are the chief collectors and repositories of spatial data used for GIS analyses. Once collected and manipulated (e.g., classified), these data must be stored. In Ontario, for instance, Mines and Natural Resources (MNR) is a data warehouse of all geoscience data. One of the problems they encounter is that spatial data are framed by time. We don't really have spatial data, but spatiotemporal entities that represent a snapshot in time. Well logs are data snapshots. As archival data continue to be collected and centralized, the problems of transforming data to suit a particular secondary project will continue to grow. Moreover, once data are available, people have to stop ignoring it. Development projects including real estate and land-fill operations in an area often don't want hydrogeological data. Developers take the view that more data are dangerous. Data do not necessarily help them achieve their objectives. With subsurface or other forms of environmental data, they can no longer say "everything will be just fine" with respect to new developments if they potentially threaten an aquifer close to the surface or lack sufficient groundwater to support a planned community. Scien-

matCode	Material description
99	miscellaneous; no obvious material code
11	covered; missing; previously bored
10	fill (incl. topsoil, waste)
9	organic
9-8	organic, topsoil
8	clay, silty clay
8-1	clay, silty clay, with rhythmic/graded bedding
8-8	clay, silty clay, topsoil
8-9	clay, silty clay, with muck, peat, wood fragments
7	silt, sandy silt, clayey silt
7-1	silt, sandy silt, clayey silt, with rhythmic/graded bedding
7-8	silt, sandy silt, clayey silt, topsoil
7-9	silt, sandy silt, clayey silt, with muck, peat, wood fragments
6	sand, silty sand
6-1	sand, silty sand, with rhythmic/graded bedding
6-8	sand, silty sand, topsoil
6-9	sand, silty sand, with muck, peat, wood fragments
5	gravel, gravelly sand
5-1	gravel, gravelly sand, with rhythmic/graded bedding
5-8	gravel, gravelly sand, topsoil
5-9	gravel, gravelly sand, with muck, peat, wood fragments
4	clay-clayey silt diamicton
4-1	clay-clayey silt diamicton, stoney
(4-2	clay-clayey silt diamicton with gr/sa/si/cl interbeds or lenses)
4-8	clay-clayey silt diamicton, topsoil
4-9	clay-clayey silt diamicton, with muck, peat, wood fragments
3	silt-sandy silt diamicton
3-1	silt-sandy silt diamicton, stoney
(3-2	silt-sandy silt diamicton with gr/sa/si/cl interbeds or lenses)
3-3	diamicton, texture unknown
3-8	silt-sandy silt diamicton, topsoil
3-9	silt-sandy silt diamicton, with muck, peat, wood fragments
2	silty sand-sand diamicton
2-1	silty sand-sand diamicton, stoney
(2-2	silty sand-sand diamicton with gr/sa/si/cl interbeds or lenses)
2-9	silty sand-sand diamicton, with muck, peat, wood fragments
*(1	rock)
1-1	limestone
1-2	shale
1-3	granite (possible bedrock, probable boulder)
1-4	dolomite
1-5	potential bedrock
1-6	sandstone
1-7	interbedded limestone/shale

Figure 3.10 *Material codes (matCode) and associated lithologies used to classify subsurface materials.*

tists at the GSC explain that data repositories have increased their responsibility to communities. The more data you collect, the more you have to explain and interpret it for developers and water managers as well as other stakeholders. Despite these challenges, standardization remains the chief vehicle for making data accessible to multiple stakeholders. A few years ago, Statistics Canada approached the GSC. They wanted to know the total volume of groundwater in Canada. This was a signal that there is a strong underlying need for data standardization. But, the question is, politically, can you make it happen?

Flexibility is the key to circumventing some of the shortcomings of the process of classification and the politics of standardization. Research in the Department of Geography at Simon Fraser University is proceeding to enable more flexible systems of classification. It uses well-log data as an instance, but is equally applicable to other spatial data sets such as roads or forests. In BC, well-log data were until recently raw. Working with Earth Scientist Dr. Diana Allen, I created a very broad classification system for well-log data that could be used across the country, and internationally, but that circumvented some of the shortcomings of iron-clad classification systems. Their first step was to allow users the option of focusing on bedrock materials or surficial materials (soils and till that overlay the bedrock). The second was to retain original descriptions while creating a standardized lithology field that allowed operators to use the data in GIS. The original descriptions are there as an archival resource for future users who question the interpretation of a particular classification system to evaluate the original terms.

Another impediment to classifying data in government agencies is that it is a laborious process that requires considerable resources. In order to automate the standardization process, my research assistant, Lee Wang, and I created a standardization application in Visual Basic. The graphical user interface (GUI) allows users to take a database of water well data from anywhere in the world – as long as it contains lithological units characterized by depth – and standardize it. The interface for the GUI is shown in Figure 3.11. The user is an active player in the process, and can choose from two systems of classification: the ORM and the more complete but complex BC classification system. The opportunity exists to add multiple classification systems that can be accessed using a pull-down menu. The user can also choose an automated or semi-automated method of standardization. In the latter, the program alerts the user to situations in which two or more attributes are possible in the context of the standardization rules. An experienced hydrogeologist, for instance, may prefer this option as it would allow him or her to make decisions dependent on the lithological context and his knowledge of the area. Users can also view the rules to determine whether they are suitable for a particular

Figure 3.11 *A graphical user interface (GUI) for an application designed to standardize well-log lithologies. The GUI allows users to automate the standardization process, or alternatively maintain control over the process by designating an equivalent term from a classification system for each lithological entry.*

Source: *First printed in* Cartography and Geographic Information Science, *vol. 29, issue 4, p. 350.*

application. A city planner using the well log to ascertain if an area is likely to be stable in the event of an earthquake might select "view rules" to determine if they favor descriptions of gravel over sand in ambiguous cases. Figure 3.12 illustrates the screen which allows the user to select options. If the rules convert the raw lithological description "sand and gravel" to gravel, then risk of vulnerability in an earthquake might be underestimated. The Flexible Standardization GUI illustrates that tools for data preparation in GIS can accomplish the pragmatic goal of classification while taking into account many of the human factors and complications associated with classification. It is an imperfect attempt to achieve standardization that accounts for human elucidation and interpretation.

Standardization and classification prepare data for use in models, but not for the politics and pragmatics that accompany their implementation. In May 2000, seven people died, and 2,300 became ill after drinking contaminated water from the town of Walkerton's municipal water

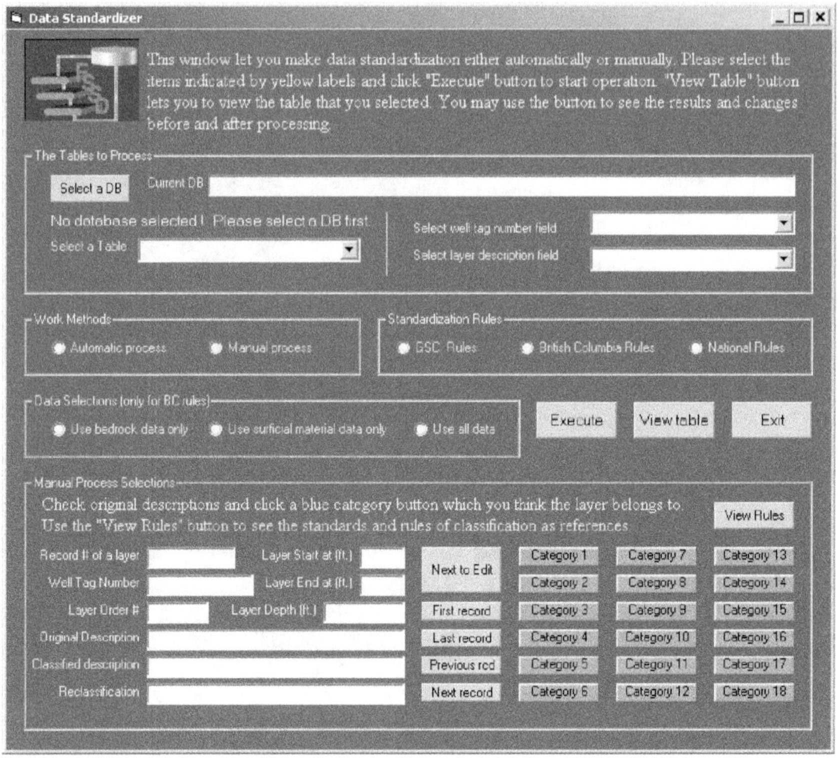

Figure 3.12 *The data standardizer screen of the Flexible Standardization GUI. Users can select a "manual" option that permits them to retain control over how each semantic term is interpreted.*

Source: *First printed in* Cartography and Geographic Information Science, *vol. 29, issue 4, p. 350.*

supply. A government inquiry published two years after the tragedy estimates that hundreds of children infected will continue to suffer life-long kidney complications as a result of the infected water. Walkerton is a small town in Southern Ontario, located on the Oak Ridges Moraine. Residents draw their water from a municipal reservoir that is fed by wells. In May 2000, the town's drinking water was contaminated with *Escherichia coli* 015:H7 (E. coli for short). The inquiry later determined that Well 5 was the source of the infection – linked to the recent spread of manure around the well followed by unseasonably heavy rains. The farming practices in themselves were not found to be the cause of the E. coli, but rather the failure of the Public Utility to adequately chlorinate the well and municipal water supply fed by it.

Leaving aside debates about the environmental implications of heavy chlorination as the preferred antidote to infections associated with manure, there is another explanation for the contamination. Well 5 was only five meters deep, and covered by unconsolidated materials known as "overburden." There were also likely conduits into the overburden such as fence posts and possibly a spring. These channels likely allowed diluted manure to travel from the surface to the bedrock, and into the aquifer. Not only were there multiple paths to the aquifer through the overburden, but the surrounding bedrock itself was fractured which permitted bacteria from the manure at the surface to quickly navigate rock channels, and enter the Well. The Honourable Dennis O'Connor who investigated the infection for the Ontario government noted that "Well 5 was the primary supply well contributing the most significant amounts of water to the distribution system."

The fragile nature of Well 5 was first noted in 1978, the year before the well was approved. At the time, however, there were no conditions associated with its approval. The Walkerton water system was subsequently inspected by the Ministry of the Environment in 1991, 1995, and 1998 but, in each instance, the susceptibility of Well 5 was overlooked. Chief GSC Hydrogeologist Dr. David Sharpe testified during the Walkerton inquiry. As initiator and head of the joint project between the Ontario MOE and the GSC to standardize the well-log data for the ORM, Dr. Sharpe was able to demonstrate the vulnerability of Well 5 based on GIS modeling of the geology. The standardization project and the GIS models that emerged were conceived more than a decade after Well 5 was approved, but the information that they conveyed had been available to the Ministry for several years. That they were not incorporated into strategic planning of water resources was due to a series of political and bureaucratic divisions that include severe cuts to the Ministry budget resulting in the loss of 750 jobs, a strong mandate in the province for privatization of resource management, and structural division between sectors of the same organization. The year 2000 Walkerton tragedy is a reminder of the importance of well-log data for environmental decision making – and protection. It is also a reminder of the politics that accompany the production and use of data and GIS analysis.

Conclusion: Data are Useful Stories about the World

Danny Dorling cleverly quipped that "data are the plural of anecdote" (2001, 1335). He is referring to the fact that data are compiled with a particular purpose in mind, and they reflect the assumptions and preconceptions of both the data collectors and data users. They are, in fact, stories about the world that change depending on the teller. This would be interesting, but tangential from a GIS perspective if data weren't the basis of all GIS analysis. The best indicator of the value of GIS is the quality of the data. But, if the results of GIS analysis are only as reliable as the data, this implies that there are good data, and bad data. Unfortunately, this is a simplistic distinction. It is more useful to talk about the appropriateness of data for a particular inquiry. The example of data collected by the Ministry of Environment, Land and Parks (MELP) and the Ministry of Forests (MOF) for crown closure and other characteristics of BC's forests offers an example of this maxim. The Ministry of the Environment's objective was to maintain Mule Deer habitat zones during the challenging winter months. The Ministry of Forests compiles data that reflect cuttable forest areas based on species, age, and size of the stands. In each case, the data were collected for the same area, but told a different story. GIS is the repository of these stories and their interpreter, but reflects their original context.

The challenge of standardization emerges when data collected to tell different stories about the world must be synchronized to produce an entirely new narrative. Water-well data, for instance, are collected to instruct property owners where to locate a well in order to minimize the cost of drilling while maximizing the yield. That data are, however, frequently the sole source of hydrogeological information used by environmental decision makers, municipal planners, and waste management professionals among others to determine where aquifers are vulnerable to human activity. In order to transform these data so that they can be used to tell a different story, they must be highly refined through standardization and classification processes. Information is lost during the makeover, but the injury to the data's integrity is generally considered desirable. The danger is not so much that data are reinterpreted, but that users of GIS recognize that all data are interpretations of the world.

There have been painful debates within geography about the epistemological context in which GIS is conducted. I have made the argument that the only way to characterize GIS is under the rubric of pragmatism (Schuurman, 2002). Pragmatism is antifoundationalist, tending instead toward regarding knowledge builders as participants rather than observers. Knowledge, in pragmatism, is a tool for organizing the world or our understanding of it. Truth, likewise, is not absolute, and can't be defined by epistemological criteria precisely because there is no outside position with which to discern it. Moreover, truth can be revised. This is as apt a description as any of the data that underlie GIS.

Derek Gregory (1994) points out that pragmatic approaches to knowledge incorporate "self-correcting inquiry." Data are often used simply because they exist, and the cost of collecting new data for a specific purpose is prohibited by the project budget. The mistake is to presume that they tell your story with the same ease that they tell others'. And the antidote is to take charge of data, ask where they came from, who collected them, who funded their collection, and how were they transformed into their present form. The need for this information is an argument for metadata, the collection of which implies recognition of the situatedness of data and the stories in GIS that result from their use. The onus is on users of GIS to engage in "self-correcting inquiry," which, in this case, implies taking responsibility for data.

4

Bringing It All Together: Using GIS to Analyze and Model Spatial Phenomena

Data and data models are important constituents of GIS, but the real power of the technology emerges from the ability to tell us more about the spatial world than is possible to discern from pieces of data stored in a computer data model. Stan Openshaw (1997), a prodigious contributor to GIS analysis techniques once quipped pessimistically that 95 percent of GISystems are used only to store data – despite their capacity for considerably more sophisticated analysis. That was certainly the case a decade ago, but early GIS data managers gradually recognized the value of using GIS to isolate and analyze spatial relationships. In this chapter, the power of spatial analysis to transform our understanding and perception of spatial relationships is explored. Discussions of data in the previous chapter emphasized that analysis depends on data; analysis also has its own logics, assumptions, and rationalities. This introduction to the query power of GIS begins with several examples so that the reader is able to imagine what separates analysis from simple data management and display using maps.

Transformations involved in spatial analysis are dependent on particular mathematical logics including reclassification and Boolean algebra. The basic tenets of these operations are explained in order to afford the reader a more thorough view of GIS operations. Spatial analysis is conducted differently in every domain, but there are common steps. Two case studies from the field of environmental management are subsequently described as a means of conveying the scope of GIS analysis and recommendations from it. Analysis is not based entirely on subterranean digital operations; much of it is accomplished through refined graphical displays that permit the user to identify relevant spatial patterns. The power of visualization as an extension of spatial analysis is portrayed with reference to a study of Tuberculosis in the Vancouver area. The final example of spatial analysis is based on an on-going population health study at Simon Fraser University (SFU). The "Urban Structures" project is described as an example of the ability of GIS to enable researchers to better understand the multiplicity of factors that affect health and well-being. Each of these examples provides insight into how GIS uses data and research assumptions to construct a version of spatial events and relationships that can be used by policy makers, health researchers, environmental protection agencies or any of a myriad organizations that rely on GIS. But GIS operates within a particular set of political and social imperatives that benefit some groups more than others. In the final section of this chapter, the *rationalities* within which GIS is conducted are discussed.

The power of spatial analysis is exemplified by cadastral systems which are developed and managed by municipalities. Cadastral systems were historically used to store survey definitions of property lots, land tenure, property values and taxation rates, roads and their condition, zoning information, utility routing, and public recreation areas. Early GIS was used to call various layers such as property lines, roadways, or common lands up on the screen for visual display. Use of GIS as a tool for managing cadastral information was quickly succeeded, however, by analysis which extended the purview of land managers. Simple queries include calculating the percentage of home owners within 1000 meters of a green space or a library. Another analysis might yield areas with greater or lesser home ownership; models can be built to assess whether factors such as income, mobility, or age of the population affect home-ownership levels. Each such query assists planners in characterizing the attributes of a community, and its access to public resources.

GIS is distinguished from cartography by this ability to analyze data. Traditional maps present a static picture of phenomena and their relationships at one point in time. It is impossible, however, to look at a single map, and ask questions like: "how has the population density changed since the last census?" Or "which areas under forest license contain old growth Douglas fir?" GIS is able to transform map data into customized information. Increasingly, the emphasis in GIS is on complex spatial modeling to support decision making for environmentalists, urban planners, police, physical scientists, and a myriad other disciplines. Many of the most powerful analysis techniques are, however, very simple. They include measurement, distance calculations, point in polygon queries, shape analysis, and slope calculation. Measurement of spatial areas is an integral GIS function. It includes calculations of perimeters, polygon areas, and line lengths. The simplest is line length calculation which is based on the distance between the two points that define the line end points. The formula for line lengths is based on the Pythagorean Theorem. For vector data, the formula uses the (x,y) coordinates for line end points. Raster distance is more easily calculated especially when the raster cells are square: the number of cells between two points is simply counted, and multiplied by the length of the cell. Distance calculations are based on line length measurements, and enable the GIS user to make informed decisions for important problems such as the distance to the nearest hospital along roads, or the length of a bicycle path between two locations. Perimeter calculations are again linked to distance calculations. They are relevant to problems that involve surrounding a spatial entity such as building a fence around the perimeter of a property or a wall around a political territory. The construction of ancient walls such as the Great Wall of China and Hadrian's Wall might have benefited from a cost analysis based on their perimeter distances involved in separating two territories.

Thus far, analysis has been treated as a metaphorical *black box* (see Chapter 1). The term refers to the hidden processes that occur at the software level in GIS. They are not transparent to users who, by implication, must trust the results. In this section, the basis for many of the operations occluded by the black box are explained, though the black box cannot be unpacked, the models, as the algorithms that populate it, are considered proprietary by GIS software companies and well-protected as a result. John Cloud (1998), a historian of GIS and remote sensing uses the expression *shutter box* to refer to the process of obtaining glimpses into the algorithms and models that provide the intelligence for GIS analysis. Clearly some technical understanding of how GIS operates is necessary to gain access to the shutter box. The following section is a primer for the basic operations that comprise GIS software.

Frequently, users of GIS want to know whether a given spatial entity is contained within an area. Point in polygon queries include: "is there an elementary school within this housing development?" Or "is there a gas station in this rural county?" An elegant algorithm for the calculation of whether a certain point lies within a specified geographical area is used in GIS. Called the "point in polygon algorithm," it tests points for inclusion based on the number of vertices between the boundary of the polygon and the point as illustrated in Figure 4.1. A more complicated form of the point in polygon algorithm uses many points and many polygons, and assigns points to a polygon. The points might be represented by houses and the polygons by municipalities. The algorithm, in this case, might assign each home to a municipality in order to assess building code restrictions applicable to individual homes.

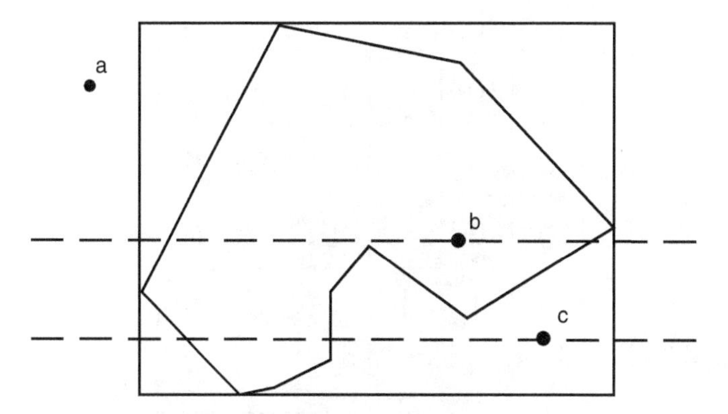

Figure 4.1 *The point in polygon algorithm tests to determine whether a given point is within or without the boundaries of a polygon. In this case, the algorithm would find that points a and c are external to the polygon while point b is within its limits. The algorithmic steps are outlined below.*

Steps:

1. *Draw a square around the polygon extent.*
2. *Compare coordinates of point with square vertices. This will eliminate many points such as a.*
3. *Draw a horizontal line through the point and the square boundary.*
4. *Count the intersections between the line and the polygon. Odd number of intersections = inside the polygon.*

Source: *Diagram adapted from Worboys, M. (1995)* GIS: A Computing Perspective. *London: Taylor and Francis.*

Simple queries can also be used to assess shape characteristics of polygons. Shape sinuosity, for instance, is the ratio between the length of the straight line between two end points and the length of the curved line as it meanders between the same two points. Such measures can be used to estimate the difficulty of river or road navigation. Shape analyses have been developed for use in wildlife habitat analysis. Michael DeMers (2000), author of one of the most thorough introductory GIS texts introduced the "edginess index" to GIS students as an example of how measurement techniques can contribute to spatial understanding and potentially to decision making. The edginess index illustrated in Figure 4.2 uses a filter or roving window that moves over the polygon being analyzed, and reassigns a value to each cell of the filter. Cells that are at least 50 percent within the polygon are reclassified as "1;" those that do not contain the attribute value are reclassed as "0." A higher number of "0"s indicates an "edgy" border. This index is used to help wildlife habitat specialists determine if a certain border will attract deer and other animals that like to graze at the perimeter of an area, but prefer a high edginess index so that they can

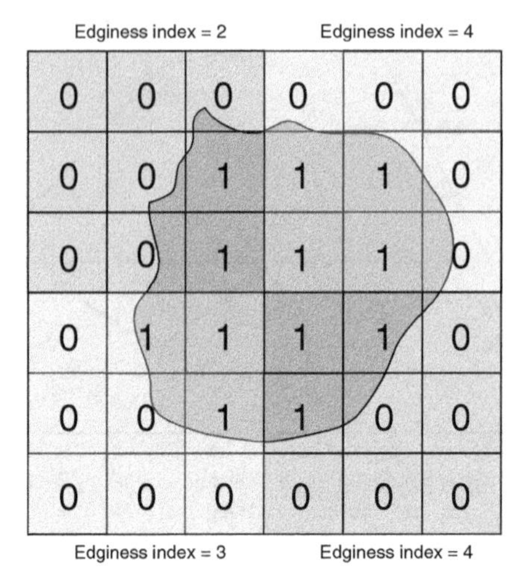

Figure 4.2 *Edginess index.*
Each of the four quadrants is associated with a different edginess index. Areas with a high number of 1s are more "edgy" as more of the polygon is connected by cells; this represents a preferable habitat for some species.

Source: *Adapted from DeMars, Michael N. (2000).* Fundamentals of Geographic Information Systems. *2nd edn. Toronto: Wiley & Sons, Inc.*

easily disappear into the forest in the case of a threat from other animals or humans. Though the edginess index comprises a simple calculation, it is an example of the utility of spatial analysis to model geographical phenomena.

As GIS developed and has been disseminated, its ability to model interactions among spatial phenomena has increased. Today GIS is used to query spatial data, analyze spatial relationships, and characterize regions, as well as to model spatial change over time and space. Spatial analysis allows the user to re-represent data so that information and new perspectives emerge. Spatial analysis techniques vary widely in complexity from simple queries and overlay to sophisticated environmental models. It is important to remember that not all spatial analysis is linked to formulae or computer queries. The eye is able to detect patterns in data through a process called "visualization." The remarkable ability of the human brain to recognize patterns is linked to the proportion of neurons in the brain that process visual data: over 70 percent of all neurons. It is this combination of visualization and computer generated modeling that allows GIS to surpass the map in its ability to communicate information. Of all forms of spatial analysis, however, overlay remains the signature GIS methodology.

Overlay Analysis, Set Theory, and Map Algebra

Overlay analysis (see Chapter 1) is perhaps the most common GIS analysis function. Its ubiquity is based on the ability of overlay to reveal areas common to two or more attributes. Overlay can be used, for instance, to find the areas of vacant land in a city that are zoned residential *and* do not cover a shallow aquifer or the number of city parks that lie within a certain residential neighborhood. Such queries may seem straightforward when dealing with several attributes over a small area, but can quickly become very complex when the number of attributes increases and the scale of analysis decreases (with an attendant increase in area). Indeed the power of GIS is precisely in the capability to efficiently analyze large volumes of data.

GIS researcher Jerome Dobson (1993) provides an example of this capacity based on a research project that would have been impossible without GIS: an investigation of factors related to continental drift. The project involved analysis of more than 160 possible attribute combinations across the globe. These combinations were easily tested by a GIS. A manual analysis would have been limited to hypotheses involving, at best, 5 percent of attributes. Dobson concludes that spatial logic of the sort enabled by GIS is suited to the "resolution of large-area, comprehensive, integrative problems." GIS enables holistic analysis of geographical problems and is differentiated from other quantitative technologies by the ability to address causal and relational questions at any scale, involving a large number of attributes (Dobson, 1993, 437).

A simple example of overlay analysis at a relatively large scale involves Spotted Owls in a park. Spotted Owls are considered an endangered species in Canada and the United States. In this example, Spotted Owl National Park is a hypothetical recreational area that contains some of the few remaining Spotted Owls in the country. The Park administrators have hired a team of students for the summer to spot Spotted Owls and record their locations. Once the locations are entered in a database, it is then possible to map a combination of factors affecting the owls' habitat. For instance, one could ask the GIS to show you where the owls were located. This might indicate what part of the Park had the most hospitable habitat for the owls. One could also direct the GIS to show what trees were in these locations and whether roads were nearby. This would indicate what types of vegetation the owls favor, and whether they are bothered by vehicular traffic. In GIS, these types of relationships are expressed as "queries." In this case, the query might be expressed as: show owl locations > 500 m from highways AND < 100 m from forest groves with the result displayed in Figure 4.3.

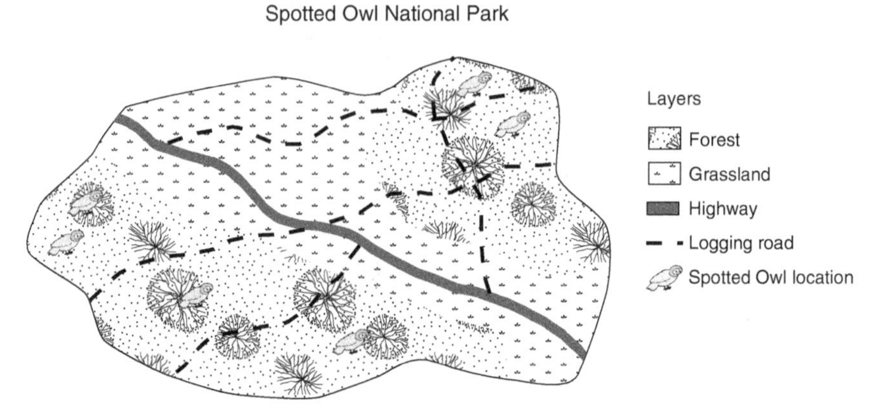

Figure 4.3 *Queries in GIS answer questions about the relationship between points, lines, and areas.*

GIS overlay analysis can be used to identify populated areas that are "at risk" for fires in Southern California. In Figure 4.4, each layer constitutes a theme that was selected based on relevance to fire risk and response. Populated areas are illustrated by dots. Rivers which help to break the spread of fires constitute another relevant layer. Road networks allow the map reader to situate her/himself; they also represent hazard response routes. A map of established fire hazard zones as well vegetation zones contributes to the visualization of the problem as well as the GIS analysis. The final map shows all populated areas within the fire zones as well as a one kilometer *buffer* zone around the fire hazard zones.

Buffers are used frequently in GIS to demarcate a zone around a spatial object which should be included in the analysis. An overlay analysis in such a case would consider not only the spatial object but its buffer. Buffers are created around salmon streams, for instance, to mark an area that should be protected from waste dumping. Noise buffers are created around highways and airports to signify land that may be clear but which should not support residential housing because of potentially hazardous noise levels. Wetlands are often buffered for the purpose of environmental analysis to protect the delicate wildlife that inhabit them. Buffers are usually consistent in width around a spatial entity, but can vary. Buffer zones created after a hazardous waste spill might vary in width to provide extra protection around a school. Likewise, a buffer around a forest might be wider around an edge area infested with Spruce Budworm as illustrated in Figure 4.5. Buffering is a way of extending overlay analysis to account for areas that are affected by spatial change as well as to designate protected zones.

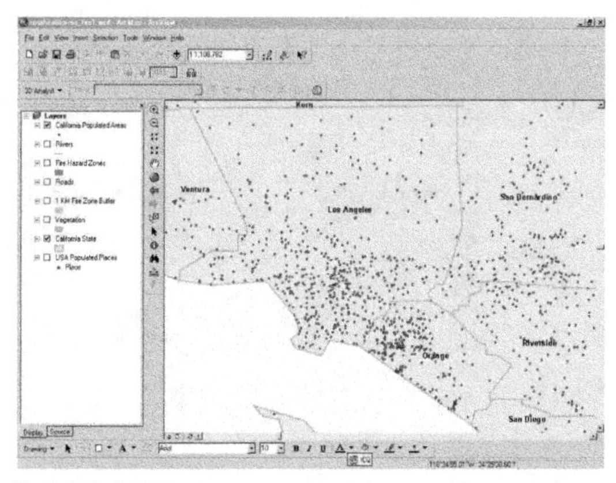

Populated areas

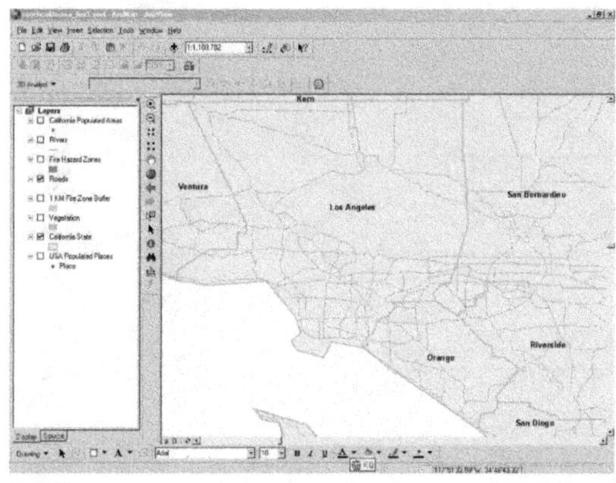

Roads

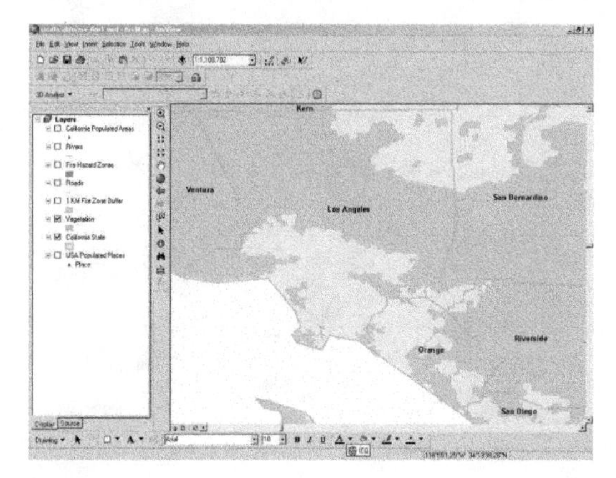

Vegetation

Figure 4.4 *Illustration of layers used to produce a map of the areas at greatest risk during a forest fire in Southern California.*

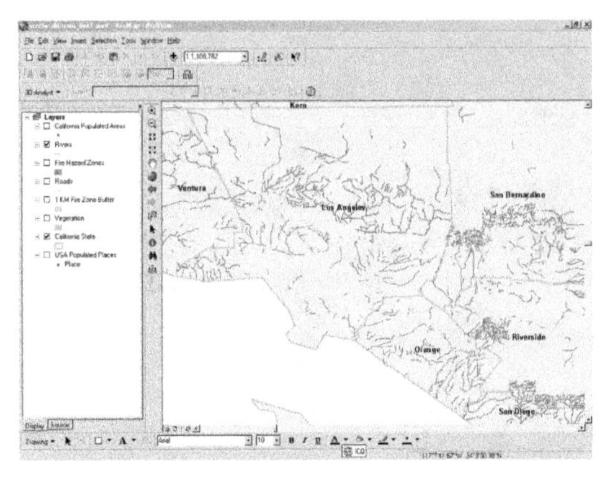

Rivers

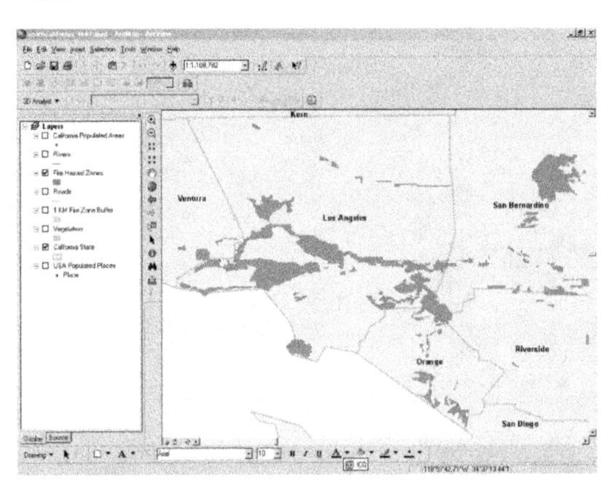

Fire hazard zones

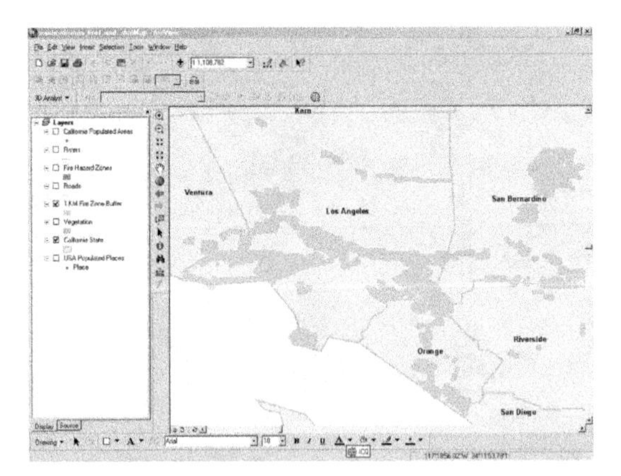

Fire zone buffer

Figure 4.4 *(Continued from previous page)*

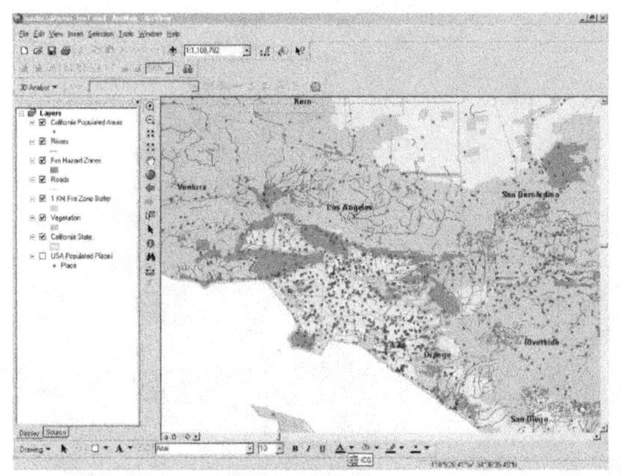

Cumulative query result showing all layers including buffers around fire zones

Figure 4.4 *(Continued from previous page)*

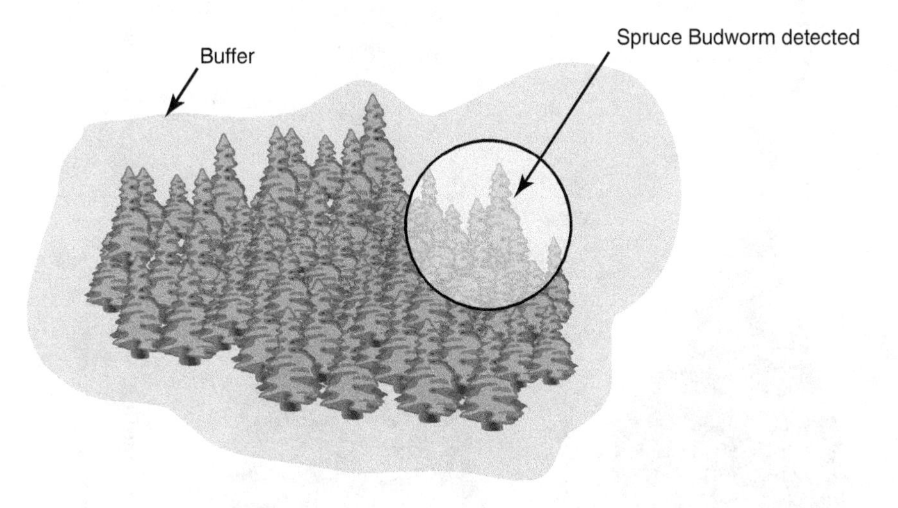

Figure 4.5 *Variable buffer.*
Note that the area where an infestation of Spruce Budworm has been detected is surrounded by a more robust buffer (illustrated in light grey).

The computational procedure for overlay is different depending on whether raster or vector data are being analyzed. Overlay is well suited to raster data, and poses little computational challenge because each layer is already registered (lined-up) spatially. In Figure 4.6 a raster coverage of the elevation surrounding Whistler Mountain, an internationally known ski resort is overlaid with remotely sensed imagery of the area as well as road networks. The overlay operation is relatively simple and elegant, and the power of GIS to illustrate spatial relationships is evident.

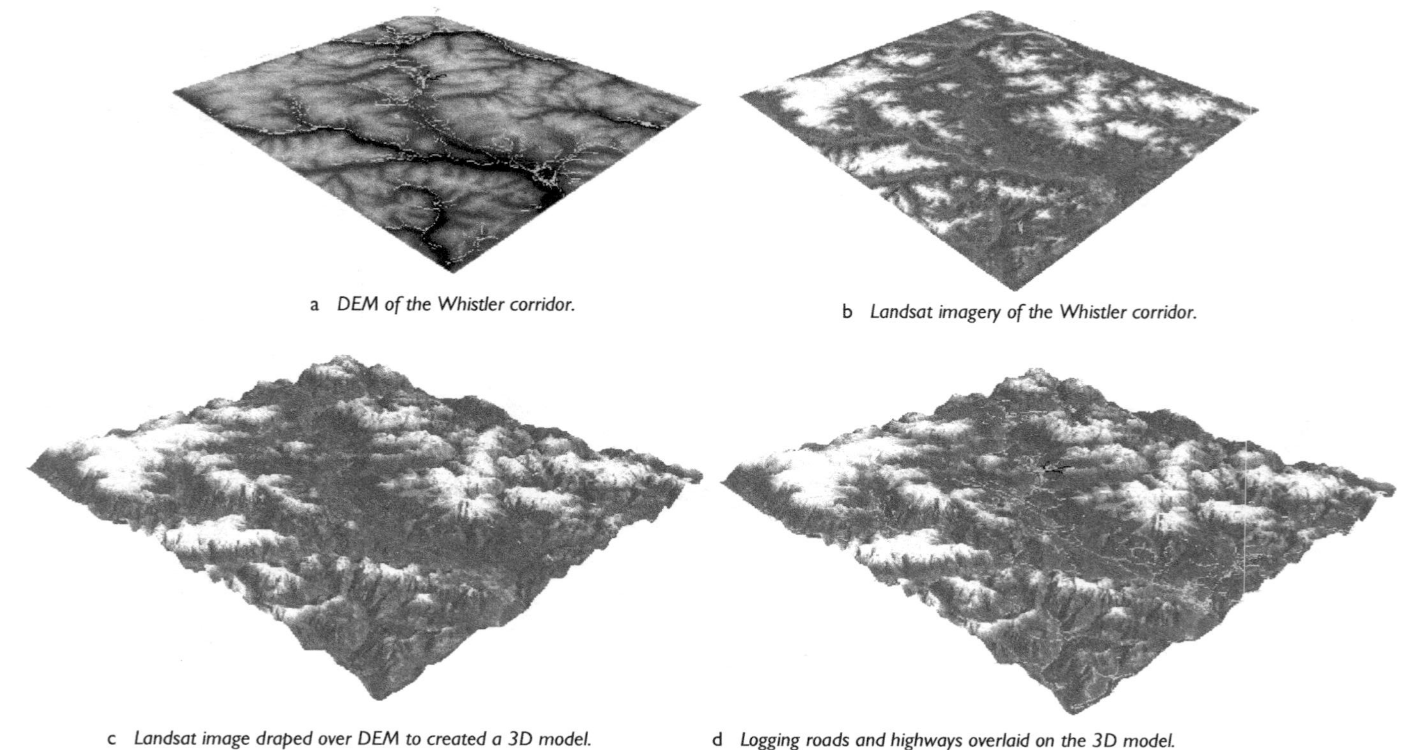

a *DEM of the Whistler corridor.*

b *Landsat imagery of the Whistler corridor.*

c *Landsat image draped over DEM to created a 3D model.*

d *Logging roads and highways overlaid on the 3D model.*
(See http : //www. blackwellpublishing.com/schuurman for color version.)

Figure 4.6 *Squamish–Whistler corridor depicted using raster overlay.*

Similarly, fallow land could be overlaid with grazing land to determine areas of potential grazing land, not presently utilized. Raster overlay in which the attribute values associated with grid cells are geographically registered is a computationally efficient way to detect pattern and relationships on the earth's surface.

Polygon overlay is a much more difficult process to implement than raster overlay. Indeed, the development of vector GIS was slowed by the difficulty in developing an algorithmic solution to polygon overlay. Nick Chrisman, a Professor of Geography at The University of Washington, recounts that it was not until the late 1970s that he and a group of colleagues working at the Harvard University Computer Graphics Laboratory were able to solve the problem by building a program called ODYSSEY. In the 1980s, commercial GIS companies were able to build basic overlay functions into their vector packages. Polygon overlay is so much more complicated than raster overlay because it entails calculating new geographies. Unlike raster overlay, in which the spatial units (cells) in each attribute layer are automatically registered, new polygons have to be calculated based on the overlap between areas. An overlay of population density data and commercial zoning involves two attribute layers, both with very different constituent polygons. Figure 4.7 illustrates that the commercial and resource zoning are associated with very differently shaped polygons than the census tracts to which population density measures are assigned. In order to glean information on where the category "high population density" overlaps with commercial zoning, new polygons need to be calculated at every polygon intersection. This process results in a proliferation of polygons to describe every area in which any of the zoning designations overlaps with each of the population density categories. After every possible polygon is described, the program displays those that are relevant to the query. The computational process is significantly more complex than that of overlaying raster attribute layers in which each square overlaps exactly.

At a computational level, overlay analysis is based on *set theory*, a mathematical construct that was developed by Georg Cantor, who published a series of papers describing his ideas between 1867 and 1871. Despite proven mathematical limitations to set theory, it is the basis for analysis in many information sciences including GIS. Set theory uses the areas or spatial entities that are the basis of GIS to formally express relationships between them. In Figure 4.8, Area A is planted with corn (A) and is related to the set of all good agricultural soil (B). In this case, the set A is contained within B. ($A \subseteq B$.) If all of B were planted with corn, then A would be described as equivalent to B ($A \equiv B$). (This presumes, of course, that corn is only planted on good agricultural soil.) Set theory allows

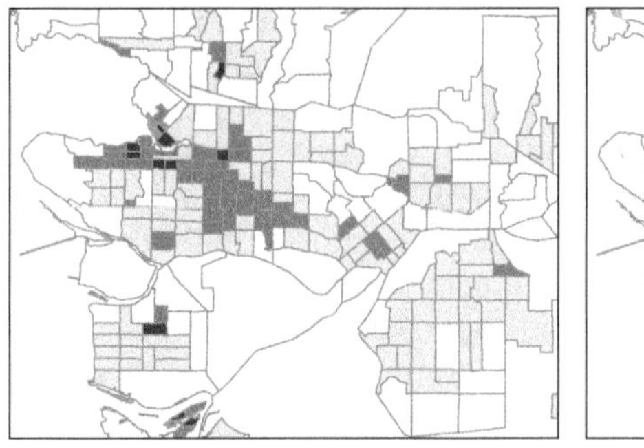

a *Population density: darker areas = higher density.*

b *Commercial/industrial zoning areas.*

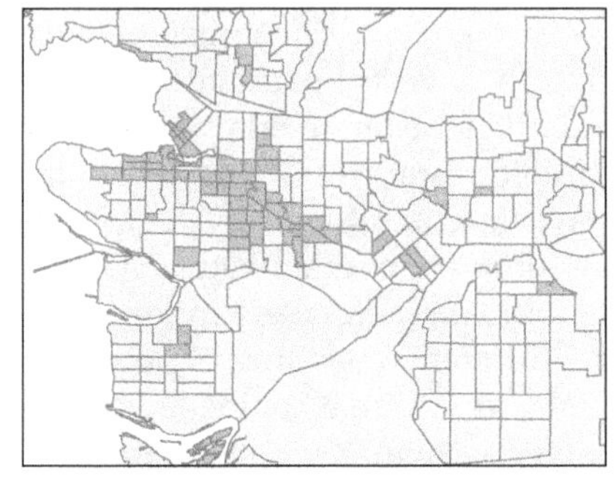

c *High population density area*

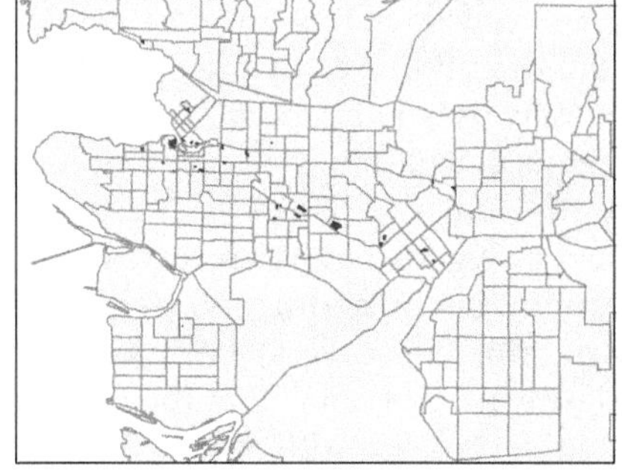

d *Commercial/industrial zones in high population density query areas.*

Figure 4.7 *Results of polygon overlay.*
Areas of high population density are illustrated in (a). Areas that are zoned commercial/industrial are indicated in (b). In (c) the areas of highest population density are isolated. These areas are overlaid with areas of commercial zoning and the output is illustrated in (d). The process of polygon overlay is more complicated computationally than raster overlay, but equally useful for discerning spatial relationships.

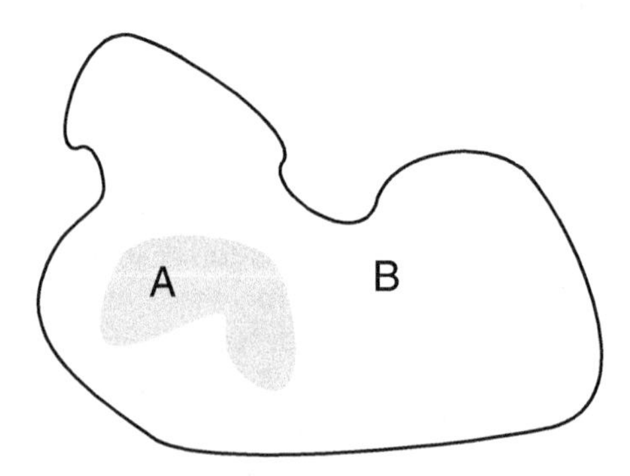

Figure 4.8 *Simple example of sets.*
The set of all good agricultural land is included in B. The area seeded with corn is contained within A. The set A is contained within the set B.

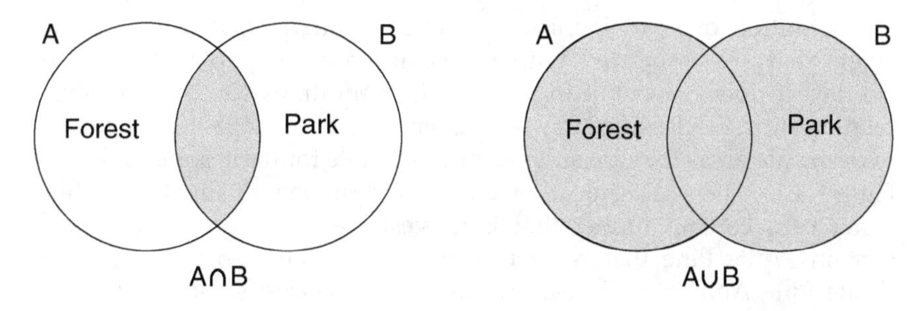

Figure 4.9 *Boolean algebra.*
If A is a forest and B is a park, then A AND B yields territory shared by both. The UNION of A and B is the total extent of the space either occupies.

logical operations to be performed on spatial entities, and is therefore of great value in returning answers to queries about relationships between spatial extents. Figure 4.9 depicts the Boolean relationships between two tracts of land: forest and park. The intersection (AND) and the UNION of forest and park are each illustrated. The operators in each case are taken from Boolean algebra. There are 16 main Boolean operators, but the most common are the ∩ (AND) and ∪ (Union). AND means the area where the two entities are both found while UNION refers to the area covered by either.

The utility of set theory is extended by map algebra, a set of arithmetical operations that can be performed on raster data. Map algebra was designed to allow raster attribute values to be transformed. Map algebra literally adds, subtracts, divides, and multiplies values associated with spatial areas. In the example in Figure 4.10, map algebra can be used to calculate a new attribute value of population density based on two input layers: area (the size of each raster) and population of each cell. A full range of operations can be used in map algebra, allowing, in effect, the reclassification of attribute values for spatial areas.

Reclassification is one of the most useful techniques because it allows the generation of new values for spatial areas without changing

Layer 1: Population

500	650	450
350	950	550
650	300	250

÷

Layer 2: Area (km^2)

25	25	25
25	25	25
25	25	25

=

Population density (pop'n/km^2)

20	26	18
14	38	22
26	12	10

Figure 4.10 *Map algebra.*
Applies the concepts of Boolean algebra to satisfy map queries. In this case, population of cell blocks divided by the areas of the grid cells to yield density. Map Algebra is a raster GIS technique that is used to calculate numerous spatial relationships.

the definition of spatial units. Like overlay analysis, it uses attribute layers that are associated with the same base geography, but unlike overlay, it does not result in new spatial definitions based on attribute relationships. Reclassification is frequently used in forest management to designate areas that are suitable or unsuitable for harvesting timber. In Figure 4.11, the reclassification query is designed to illustrate which raster cells contain timber that is harvestable based on whether they contain White Pine that is over 40 years old. The query expressed as White Pine AND >= 40 years results in the reclassification of the two input layers.

Reclassification does not necessarily depend on multiple input layers. It can be used to create categories based on suitability or as a data simplification exercise. For instance, one map layer might contain an attribute value for the number of persons with tuberculosis estimated to live in each area. A reclassification would designate each of the areas in terms of their qualification for government resources for treatment and prevention. Figure 4.12 illustrates the conceptual basis for such reclassification using raster cells. Reclassification can also be used to simplify attribute data that contain too many values for users to understand.

Layer 1: Age of trees

15	12	17	29
12	20	30	42
20	25	42	53
27	40	40	65

Layer 2: Type of trees

WH	WH	WH	DF
WH	WH	DF	DF
WH	DF	DF	WP
WP	WP	DF	WP

DF = Douglas Fir
WP = White Pine
WH = Western Hemlock

Query: White Pine AND >= 40 years old

0	0	0	0
0	0	0	0
0	0	0	1
0	1	0	1

1 = True
0 = False

Figure 4.11 *Reclassification of raster values.*
In this example, areas suitable for logging – or candidates for environmental protection – are calculated by comparing attribute values of tree age and species for the spatial cells and creating a new layer that is reclassified as White Pine over 40 years old.

Layer: Population with tuberculosis

2	0	0	5
0	15	1	0
0	5	0	1
0	10	3	25

Query: Areas >= 10 (qualify for Government resources)

0	0	0	0
0	1	0	0
0	0	0	0
0	1	0	1

1 = Qualified
2 = Not qualified

Figure 4.12 *Simple reclassification used as a policy tool.*
In this example, cells with a high incidence of tuberculosis are reclassified as eligible or ineligible for extra medical resources.

A good example of heterogeneous data values is offered by housing values in a given area. Values might range from $292,000 to $567,500 over even a small area. A larger area would likely show a wider spread. Reclassification is useful in this case to divide values into categories such as: economical, affordable, and luxury.

Set theory and map algebra are the basis for much modeling and analysis that are developed using GIS. They underlie the structured queries including overlay analysis that planners and modelers use to develop decision-making scenarios and to predict spatial change. The following sections will present three examples of spatial analysis for solving specific geographical problems: (i) a model that was developed to assess environmental pollution of industrial plants; (ii) a location analysis for creating a landfill for urban waste; and (iii) an example of explanatory data analysis using visualizations. These problems illustrate common methods of GIS analysis based on reclassification, set theory and map algebra. Moreover, they demonstrate the potential of GIS to assist in spatial decision-making.

Spatial Analysis in the Field: Environmental Modeling

Fang Chen and Julie Delaney (1998; 1999), both at the University of Western Australia, developed a model for determining the levels of industrial pollution associated with industrial parks. The model takes into account the type of industry and its known pollutants, and draws on both GIS and pollution modeling tools. Rather than use only GIS or environmental modeling tools, the project integrated the two. For example, noise quality models developed by environmental modelers were linked to GIS-based locational data on noise emitters as well as terrain information for estimating noise levels. Likewise, air quality modeling tools based on plume dispersion techniques were linked to GIS. The spatial database, locational information of pollution sources and definition of subject areas were stored in GIS, while the modeling of pollutants was done using modeling software. The results were then exported back to GIS for review and analysis. The schema for this integration between GIS and environmental modeling tools is illustrated in Figure 4.13.

The integration project begins with spatial data representation and organization in GIS. Intermediate processes analyze environmental impacts including air emission, noise and risks using specialized quantitative tools developed by environmental modelers. These are then imported back into GIS and analyzed using overlay techniques. By creating a connection between existing modeling environments and GIS, the researchers were able to extend the capacity of GIS to assist in decision making about the environmental impact of tenants in industrial parks.

Hypothetical tenants were devised for the industrial parks, and their pollutant levels estimated based on known outputs from existing plants.

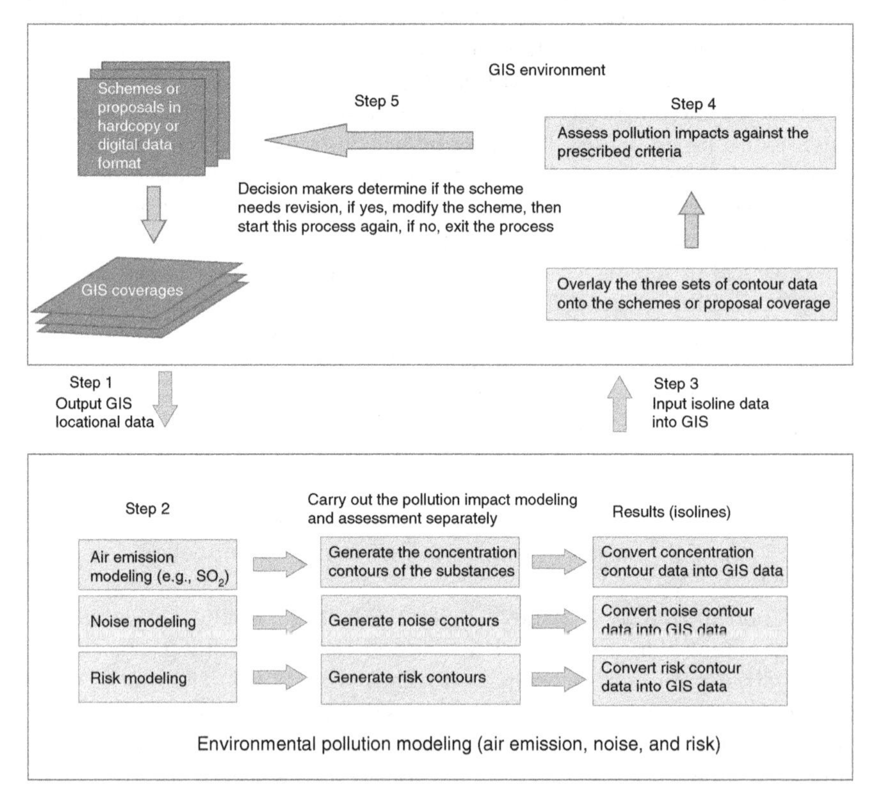

Figure 4.13 *This flowchart illustrates the links between GIS and environmental modeling environments that enabled Chen and Delaney (1999) to develop a sophisticated model of air and noise pollution.*

Based on templates of pollution associated with different industries, environmental models of noise and air pollution were generated. For East Rockingham Industrial Park in Western Australia, a hypothetical steel mill, pigment plant (using titanium dioxide), and a chemical plant were tested for their potential environmental impact. The steel mill data included plans for the mill, its levels of SO_2 (sulphur dioxide) emission and point sources for the pollution within the plan layout. These were analyzed using an existing air pollution model that input locational data from GIS. Figure 4.14a illustrates the estimated ground level concentrations of SO_2 using contour lines, while Figure 4.14b is an isopleth map of noise levels surrounding the park. Analysis of the estimated SO_2 outputs found them to be below limits sets by the Environmental Protection Agency. Likewise, noise pollution levels are shown using contour lines. In this case, the 35 decibel (dB) noise contour line remains within the limits of the industrial park, and is thus permissible unless the noise of other tenants is factored in. Figure 4.15 illustrates that the composite risk factor that was developed to estimate the viability of the steel mill. In this case, the risk contours are mostly contained within the bounds of the

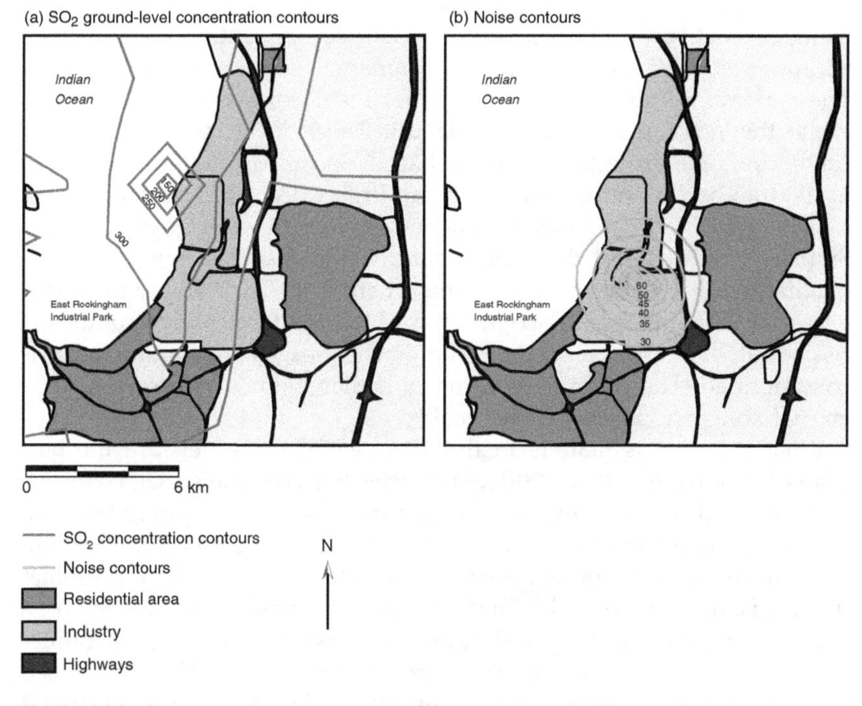

Figure 4.14 *Environmental models created by Chen and Delaney (1999) using GIS and environmental modeling tools.*

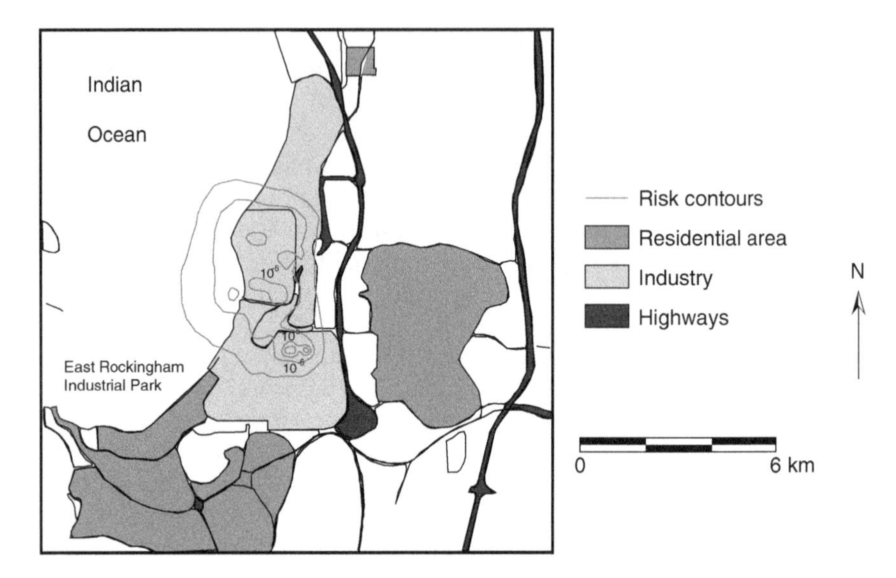

Figure 4.15 *Risk contours for development of steel mill.*

contiguous industrial parks. An important component of such analyses is, however, how the levels of environmental impacts are determined. The maximum hourly output of SO_2, for instance, was set at 350 $\mu g/m^3$ while the noise maximum was 35 dB. These levels were set by the Environmental Protection Agency based on *models* of risk related to individual health and the environment. Proscribed levels may represent a higher number than nongovernmental environmental groups advocate. Similarly, residents of the nearby subdivision may not agree with the maximum noise pollution level being set at 35 dB. Their ability to experience everyday life may be compromised by this level. When composite risk profiles are assembled, the contingency of each contributing factor is compounded. The question becomes whether models are useful if they are not able to accurately reflect reality.

What a model is matters in this discussion, as models are the best example we have of the interdependencies between the social, cognitive and technical realms. Models rely on a morphism or mapping between the entity and the representation. In this example the entity is the amount of environmental pollution generated by a steel mill, and the representation the isolines associated with risk levels. Of course, if we were to take a stroll around East Rockingham Industrial Park, we would not encounter any isolines. They are imaginary lines that represent total compounded air and noise pollution. During our walk we might find, however, that it is too noisy to talk and that our throats become a little scratchy from particulates in the air. The relationship between the lines and our symptoms is relevant, however, as models do not pretend to be the real world but to determine critical properties of a system.

Translating geographical questions into variables in order to analyze the relationship of those variables to each other is part of the shift in GIS toward decision making and prediction. Simulation models are usually preceded by a flowchart of various risks at a given location, and for its environs. A series of steps, used to derive the final sum of risk values, is charted diagrammatically. These are then formalized in a GIS or environmental modeling program. It is important to stress that authors of such systems are usually aware of the limitations of models. Many avalanche models, for instance, operate best with complete historical record of avalanche frequency, and do not make any claims for the model other than its use for assessing generalized risk. Indeed, they explicitly state that the process of simulation generates an artificial history in order to draw inference (Keylock, McClung, and Magnusson, 1999). Nevertheless, models are the basis of the power of GIS to assist in environmental decision making. Models begin as theories, but become programs (Schuurman, 1999). Once a model is ensconced in GIS, it manufactures entities. In the case of the steel mill, the entities produced are areas associated with unacceptable environmental risk.

Manufactured entity relationships substitute for a broader understanding of a reality that we cannot apprehend – or model. Each modeling system generates abstract entities (like isolines) and these lines are a form of virtual reality, a rendering. This doesn't imply that rendering physical and social environments is pointless. Rather, rendering allows us to find rules for predicting the outcome of experiments in that environment, whether it be the physical area surrounding an industrial park or the rules used by a GIS to predict increases in sea-level. Nor does this discount the possibility of realism – only direct experience or representation of the world. In this context, it doesn't make sense to talk about the right model or the wrong model; it is more useful to be guided by the axiom that "all models are wrong, but some are useful." The strength of Chen and Delaney's model is in its ability to integrate spatial and non-spatial data to assess environmental pollution. The result is a far more powerful set of tools than currently exists independently in GIS or in the field of environmental modeling. The key is the combination of *spatial* analysis with techniques for estimating industrial pollution.

Building Intuitive Models: Multi-Criteria Evaluation

One of the most common applications of GIS is to aid in decision making for spatial location. Multi-criteria evaluation (MCE) is a raster based modeling tool that allows users to combine several criteria (attributes) in order to derive a suitability index for location of a spatial entity. The first step in MCE is to define the problem and relevant criteria. Then each criteria is scored depending on its relevance to the spatial solution. The scoring remains a subjective process; its strength is in the ability to use scoring criteria commensurate with the goals of the analysis. High levels of relevance are associated with higher weights which are, in turn, used to enhance the impact of that particular factor in the suitability equation. The value of MCE is that it allows the user to weight numerous criteria in order to fine-tune the model. Moreover, conflicting criteria can be used, allowing the model to incorporate more than one point of view, and still provide results. MCE is used to assess suitability of a particular location for a particular purpose. The Chen and Delaney model illustrated the potential environmental effects of locating a steel mill at a particular location. MCE, by contrast, might be used to find the least offensive location for a steel mill, based on a series of criteria. MCE is used by municipalities, for instance, to locate services as well as sites for environmentally stressful operations.

The Greater Vancouver Regional District (GVRD) consists of 22 municipalities in the Vancouver region. The role of the GVRD is to provide essential services including water delivery, sewage treatment, and garbage disposal for the communities in its jurisdiction. Sharing services allows for greater equality among residents in the area, as well as economies of scale with respect to delivery of services. The municipalities in the GVRD share landfill facilities for solid waste. Since 1989, much of the area's waste has been transported to Cache Creek, a 48 hectare site adjacent to the Trans-Canada Highway in a community several hundred kilometers north-east of Vancouver. In the first eight years that the landfill existed, approximately 300,000 tones of GVRD garbage were deposited there. Since 1998, the levels of waste have more than doubled in response to the closure of a local landfill. At the present rate of waste disposal from the regions' two million residents, the Cache Creek site will be at capacity in 2007. In 1995, the GVRD created the "Solid Waste Management Plan" which includes strategies to reduce garbage and encourage recycling. Despite these initiatives, the need for expanded landfill facilities was writ large and in 2000 the GVRD bought Ashcroft Ranch, a 4,200 hectare site proximate to Cache Creek. Only 200 hectares of the Ashcroft Ranch site will be developed for landfill. The difficulty was to locate the most suitable area of the Ranch based on multiple variables related to environment protection, proximity to facilities at the existing site as well as roads, archeological sites, and slope.

Locating a landfill is a contentious issue with environmental, economic, and political restrictions in addition to geographical considerations. Below MCE is illustrated with respect to the location of the optimal site within Ashcroft Ranch for the GVRD's new landfill. The model created here is only for the purpose of illustration, and does not reflect any part of the process that the GVRD may have used to locate the site. The MCE model described here and the final site chosen by the GVRD are, however, very close. MCE is based on the use of both factors and constraints. Constraints and factors are both criteria. A constraint is a Boolean criteria (yes or no) that limits analysis to particular geographical areas. Factors are criteria that define the degree to which a region is suitable. Factors are usually expressed as scores. In other words, factors are conditions that influence the suitability of a given piece of land while a constraint is a condition that limits the suitability of alternatives. A modified list of constraints includes meeting government requirements for landfill location, limiting population exposure to the smell and air pollution associated with landfills, and managing public fears of unsightly land. Factors include compliance to access to roads, distance from water, distance from First Nations archeological sites, proximity to supplies of material used to cover the face of the landfill, slope, and geology (see Table 4.1).

This list of factors and constraints, like many developed for MCE is not associated with absolute, objective values. Rather MCE accommodates multiple perspectives, and the results of analysis vary depending on how factors and constraints are ranked *in relation to each other*. After the weighted model is developed, the final step is to combine all the information in order to develop a *composite index of suitability* and use it to locate the optimal site for the new landfill. The results of the MCE analysis are illustrated in Figure 4.16.

Is this the best site for the landfill? It depends on who did the MCE, the data quality and the selection of criterion and factors. To some extent, models always reflect agenda. Brian Harley, a famous scholar and critic of cartography once quipped that "the map is not the territory"(1989, 233). Likewise models are not the territory but a way of simplifying representation so that we can better interpret the viability of environments for specific applications. The problem with models is that they are confused with reality when, in fact, they are complex systems subject to change with modification of every factor and constraint. Kristin Shrader-Frechette (2000) illustrates the potential for models to be differently interpreted depending on the motivations of policy makers. Hydrogeological models are used as the basis for locating burial sites for nuclear waste. These models are in turn being used by the United States Department of Energy (DOE) to assess the suitability of Yucca Flats, Nevada for the burial of high-level radioactive waste. One panel of experts in 1992 used such models to determine that the site is not well suited for long-term waste burial given projections of tectonic activity, and other uncertainties. A subsequent report in 1995 found that the geological record

Table 4.1. *Criteria for the location of a landfill for the Greater Vancouver Regional District within the 4,200 hectare Ashcroft Ranch*

Criteria	*Factor or constraint*	*Condition*
Government regulations	Constraint	Minimum 100 m from any surface water; minimum 300 m from any dwelling site; and minimum 15 to 50 m from property boundary.
Site specific considerations	Constraint	Minimum 100 m from known archeological sites; minimum 300 m from currently irrigated land (assumed to be under cultivation); slopes over 40 percent should be excluded.
Proximity to roads	Factor	It would be preferable to locate the site close to existing roads.
Proximity to existing buildings	Factor	It would be preferable to locate landfill sites as far away from existing buildings as possible.
Proximity to archeological sites	Factor	It is desirable to locate the landfill sites away from known archeological sites.
Proximity to surface water	Factor	To reduce potential environmental problems, site should be not be proximate to water (lakes, streams, etc.).
Land cover	Factor	It is preferable that the landfill be located in open areas, rather than wooded areas.
Slope	Factor	Slope of the landfill should neither be flat, nor overly steep.
Geology	Factor	The landfill should be located on geological material that has low permeability such as clays or unfractured bedrock with a corresponding low water table.

supports the burial of the same waste at the same site. Clearly, the stakes are high in the interpretation of scientific models of earth systems, and their relationship to the world can be differently interpreted. In the case of the GVRD's landfill location, the model predicts the best site given those criteria and their modifications. Another set of criteria would likely result in the selection of a different site. MCE remains a powerful tool for GIS, however, as it allows disparate groups with different requirements to negotiate both factors and constraints. It is a rational tool that accommodates an irrational process.

The Power of the Eye: Visualization and the New Cartography

The projects described above illustrate the utility of structured spatial analysis, but they fail to emphasize the power of the human eye in

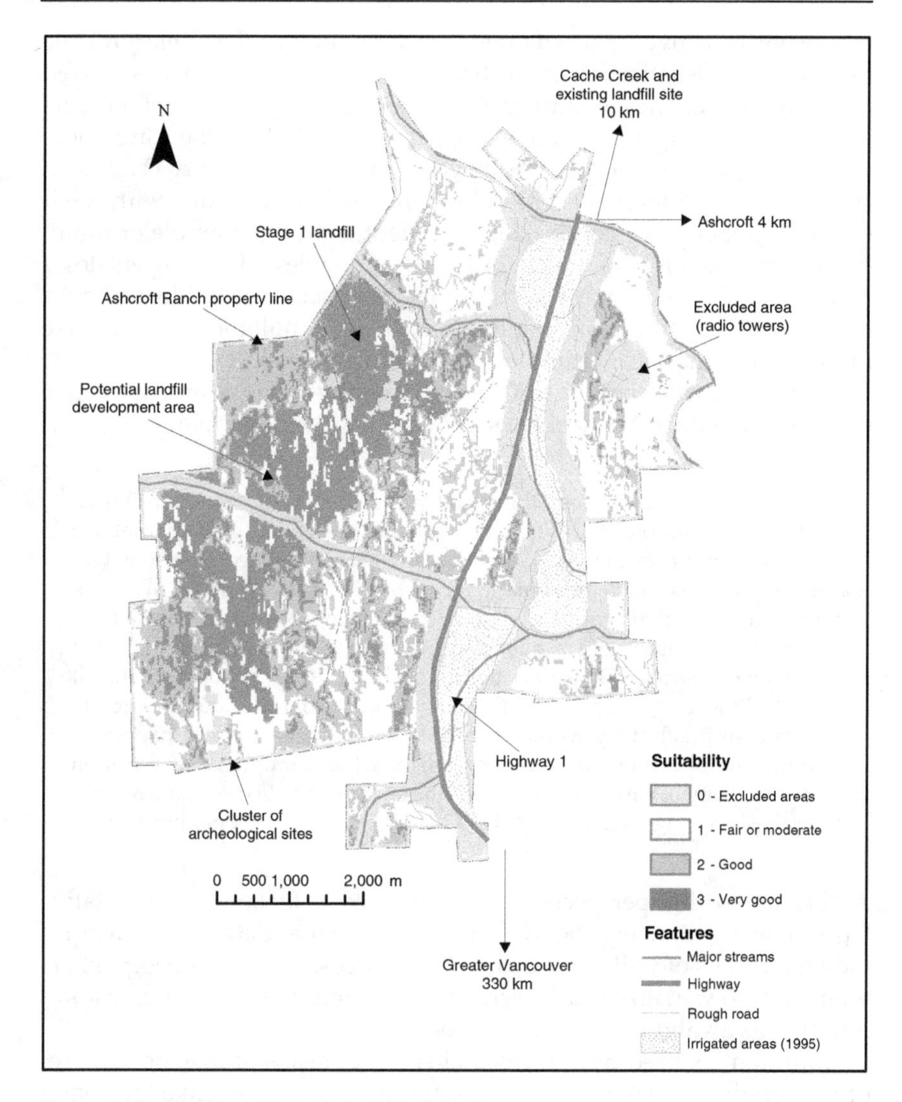

Figure 4.16 *Landfill suitability with the Ashcroft Ranch.*

Source: *BC TRM Data. Assessment of Resource Potential Ashcroft Ranch: vols 1 and 2. Golder Associates Ltd (1999).*

detecting pattern, and the role of subjectivity in GIS. Quantitative or structured inquires are certainly integral to GIS, but visuality and "intuition" differentiate GIS from the approach to geography that dominated the quantitative revolution. From the early days of GIS, there has been a divide between people using the computer to *analyze* spatial data and those using to *map* it in graphical form. The latter relied on the power of the visual to convey patterns and concepts. Rather than interpret the trend toward visualization as an abnegation of quantification, it is viewed by many as a distinct advantage. GIS becomes a way of "demathematizing" or extending quantitative approaches such that they are more exploratory, but associated with fewer measures of certainty (T. Poiker, 1997, personal interview). GIS allows researchers to work with a far greater number of variables. This is evident from the example of multi-criteria evaluation above. As one adds variables, the analysis loses mathematical precision. *Measures* of precision, such as confidence limits, become difficult to establish and attain. Instead multiple variables, like those used in the MCE example above, are permitted.

In characterizing the role of GIS in promoting intuitive and subjective exploration of data, Michael Goodchild makes the point that:

> [GIS] has reinvigorated something that was in danger of being moribund. To take an example, the kinds of methods of spatial analysis that we were developing in the late 70s and early 80s represented by geomathematical analysis were becoming very abstract and abstruse. We propounded the notion that because they were being published they would eventually be used, but realistically there was no prospect that that would ever happen. GIS came along and initially the notion was that it would allow us to implement those methods and make them easier to use, and so finally they would be usable. In practice what has happened is quite the opposite. GIS has reestablished the importance of intuition and simplicity of exploration over those very hard-core confirmatory hypothesis-testing techniques. (M.F. Goodchild, 1998, personal interview)

According to this perspective, GIS extended a lifeline to quantitative techniques by allowing the visualization of spatial data as well as providing a means of utilizing various data sources. It presents geographers with ways to visualize spatial arrangements and, in the process, restores intuition as a valid heuristic technique.

"Intuition" is used by GIS researchers as a means of making sense of or interpreting visual displays of geographical data. It is linked to cognitive processes that remain poorly understood. Links between visual cognition, knowledge discovery, and computer technology have been the subject of intense research during the past decade partly because visualization represents a radical departure from traditional decision-making processes, for many geographers. Researchers in "scientific" visualization stress that it is the relation of graphical display to communication of information that distinguishes the methodology.

Exploratory data analysis (EDA) and knowledge discovery in databases (KDD) are related methodologies that rely upon assumptions of direct correlation between visualization and communication of information. These pursuits are made possible through the convergence of digital data sets and algorithms for pattern discernment and computing. KDD and EDA comb through existing data to judge their suitability for a given analysis. Often referred to as *data mining,* KDD incorporates machine-learning computational techniques with visualization in order to perceive patterns and to judge the suitability of data according to established criteria. KDD and ETA are iterations of what is known as *scientific visualization* – a combination of simulation, data analysis, and visualization of complex scientific relationships such as chromosomal structures and ecological interdependencies. Application domains include medical imaging, genetics, biochemistry, ecology, and atmospheric science. In GIS, geographical visualization is emerging as a subspecialty that focuses on how humans interpret visual imagery, algorithms for data manipulation, and patterns of human–computer interaction. Like KDD, geographic visualization is a tool used in manufacturing meaning from data. Ironically GIS and cartography have always been based on the principles that recently brought scientific visualization to the fore. Figure 4.17 illustrates a simple visualization of the incidence of tuberculosis in the Greater Vancouver area. Spikes in incidence are clearly indicated by the height of the surface and the change in the color palette. This is a powerful technique for understanding spatial phenomena.

In research at Simon Fraser University, Dr. Suzana Dragicevic and I have collaborated with Dr. Mark Fitzgerald, Head of the Centre for Clinical Epidemiology at the University of British Columbia to

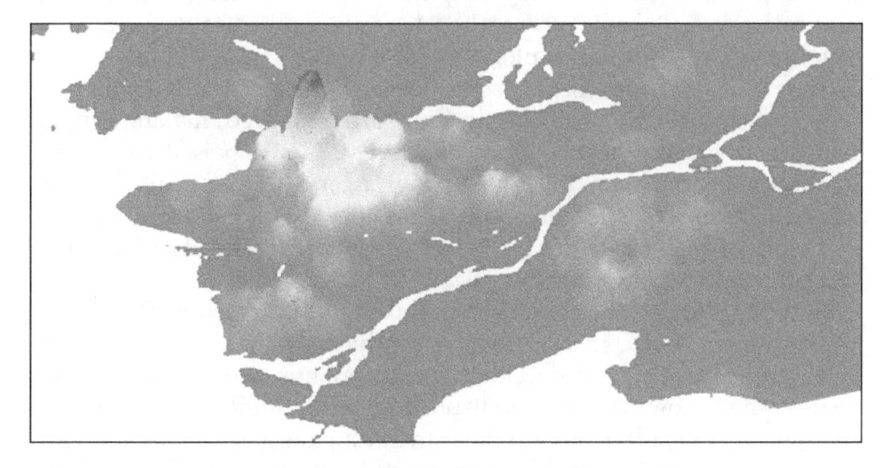

Figure 4.17 *Visualization of tuberculosis in the Greater Vancouver area.*
The spike in TB incidence occurs in an area of the Downtown that includes Chinatown and Stratchona, both areas that present a depressed socioeconomic profile, high levels of immigrants as well as aboriginals.

investigate the role of EDA in controlling infectious disease. This study (2003) developed an exploratory data analysis procedure to improve understanding of demographic factors related to Tuberculosis incidences in the Vancouver region. Tuberculosis (TB) consists of a group of pathogens that reproduce slowly leading to epidemics that can last centuries. Today there are approximately nine million cases of TB worldwide with incidence rates in any one place usually measured per 100,000. People with active TB, including multidrug resistant TB, are more mobile than ever before. Right now, 10 percent of TB cases are drug-resistant, and immigration is a factor in the spread of these strains. The highest levels of drug resistant TB worldwide are in the western reaches of the former Soviet Union followed by areas of China, and then India and Pakistan. There is a political economy related to the spread of TB as many of the patents for TB drugs have expired making diseases like HIV/AIDS more attractive to research and treat. Given that prevention of TB is more cost-effective than treatment, it makes sense to develop strategies to identify at-risk geographical areas and allocate resources for TB control accordingly. The goal of this study was to develop more sensitive, encompassing tools for identifying areas in Vancouver at risk for spread of TB.

Two main paradigms have dominated studies of the spread of infectious diseases such as TB. The two approaches arise from different investigative methods used by medical scientists and public health researchers. Doctors and epidemiologists are generally concerned with individual behavior while those who tend to work with sick people in their neighborhoods are more likely to focus on the well-being of entire communities. This has resulted in epidemiologists tending to focus on a confirmatory or "predictive" approach while public health researchers have generally used an exploratory or "causal" mode of inquiry of investigation and analysis. A consequence of this division is that resources and expertise have become compartmentalized, making it more difficult to address health issues effectively. GIS is poised to make a significant contribution to epidemiological and public health studies precisely because it can combine confirmatory (predictive) and exploratory types of analysis. In this study, we used individual and social factors that contribute to the spread of TB.

Ethnicity is one of the social factors linked to the spread of TB; it is not a determinant, but a marker for the confluence of a variety of factors that accelerate TB spread. Other socioeconomic indicators included homelessness, low income, predisposition to the disease, gender, age, aboriginal status, place of birth other than Canada, and year of immigration. Individual traits were based on the molecular epidemiological strain associated with each case of TB. One of the greatest advances in TB detection is linked to the molecular epidemiological characterization of particular strains of TB. The clustering and spread of each strain is variably linked to the socioeconomic factors listed above in addition to HIV status. Any visualization or statistical technique for analyzing TB needs to differentiate strains in order to make links with these factors. Otherwise possible clustering may be irrelevant as clusters might include multiple strains which are clearly not related to the same diffusion factors.

In this study, the analysis capabilities of GIS were combined with visualization facilities to provide a means of implementing EDA in an epidemiology context. The EDA process consists of the following stages:

1. exploratory visualization using a GIS;
2. statistical and numerical analysis of spatial patterns and trends (factor and cluster analysis in this case);
3. visualization and communication of results using a GIS;
4. repetition of stage 2 if needed with different statistical and/or numerical methods;
5. final results as tables and graphs, maps of disease clusters or scenarios of disease propagation.

By using spatial analysis techniques to identify clusters, and then displaying those patterns, the researchers were able to able to identify five major clusters of TB in the Vancouver area as illustrated in Figure 4.18. One of the shortcomings of the study was that the data included a number of missing values. In order to ensure consistency and integrity in the results, the study was forced to discard instances in which attributes were not included. Moreover, the data were not collected specifically for spatial analysis; postal codes rather than addresses were used to identify individual cases. As a result, analysis linked to the spatiality of the data was limited, and some clusters may have been missed. These factors point to the need for continued cooperation between GIS professionals and medical epidemiologists in order to design data collection and standardization.

Visualization of the results was affected by data reduction, but also by privacy constraints. An 800 m offset for each data cluster was chosen to protect the privacy of individuals. This means that map users may be misled about specific local factors related to diffusion of disease. This is a necessary compromise, however, in working with individual health data. Despite these constraints, clustering is evident based on the EDA. Moreover, the use of spatial analysis to determine relevant cofactors enhances the accuracy of EDA, and moves GIS from a visualization technique based solely on intuition to one grounded in a more structured form of analysis.

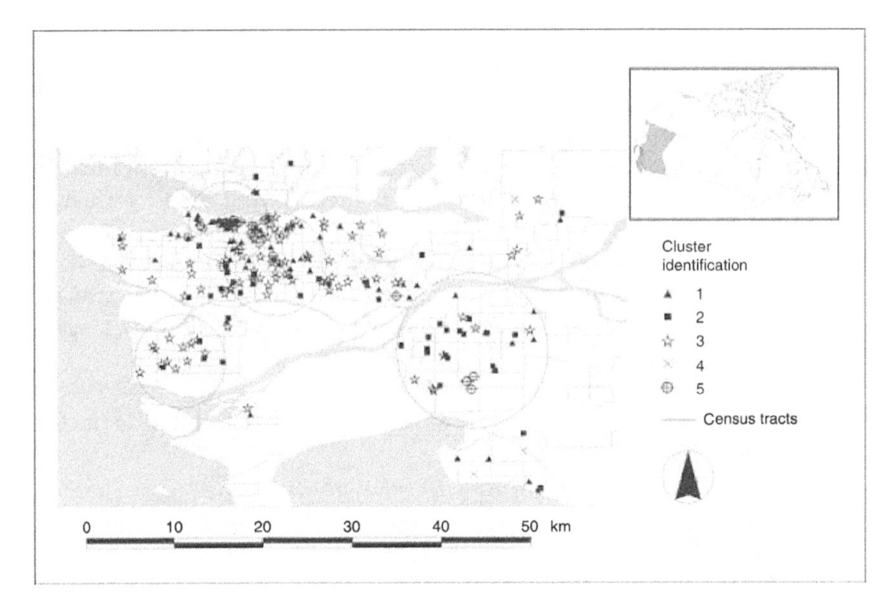

Figure 4.18 *Clusters of tuberculosis.*
Exploratory data analysis (EDA) is a structured form of geographical visualization that combines numerical and representational techniques. In this instance, five distinct clusters of tuberculosis were identified.

From Data to Analysis: A Case Study of Population Health

It should be evident that data and analysis are closely linked. This is especially true for epidemiological and population health studies which are often constrained by availability of data, their quality, and constraints to their integration. A large population health study underway at Simon Fraser University's Institute for Health Research and Education (IHRE) is led by Dr. Mike Hayes with Dr. Suzana Dragicevic and me serving as the GIS experts. As is typical in such projects, the vast majority of the work was done by Darrin Grund, the Institute's GIS analyst.

Population health is concerned with the influences that social relations play in shaping the health of individuals and communities. There is a strong tradition of biomedical focus in health research with emphasis on individual characteristics (risk factors) and disease-specific processes. Social processes receive less attention, but exert a considerable influence on health of individuals and whole communities. Housing is a good example. There is relatively little work to date on social, economic, and cultural characteristics of housing in relation to health, but housing surpasses the mere provision of shelter and privacy. It facilitates the experiences of everyday life and helps stabilize the individual with respect to the world; it is fundamentally bound up in one's sense of control over life circumstances. Housing markets also play a significant role in the distribution of wealth. Yet, in spite of this profound influence, there is a dearth of work on the health consequences of inequalities generated through housing. Our study set out, therefore, to integrate multiple factors including housing, income, crime rates, environmental pollution, and green space into an analysis of population health status in the Greater Vancouver Regional District (GVRD).

One of the first goals of our working group was to create an integrated data base for the GVRD, combining information from the census with data sources from health regions, municipalities, real estate boards, vital statistics, police departments, and miscellaneous data sets available throughout the region, as well as with the BC linked data base (containing personal health data linked to the postal code level). Two major integration challenges present themselves: (i) spatial boundary definition; and (ii) scale definition.

Spatial boundary definition presents a challenge because data from municipal, regional, and provincial policy makers are all associated with different geographic regions. This is a common difficulty in working with vector data, and is one of the chief reasons that researchers typically limit themselves to one type of data set associated with a specific set of boundary definitions. Population health researchers, for instance, may choose to use census data to develop profiles of neighborhoods, and, therefore, link any supplementary data collection to enumeration areas. This is a limiting approach, however, as it assumes that spatial definition (e.g., enumeration areas) appropriate for collecting socioeconomic data at a national level are equally suitable for understanding population health at the neighborhood level. Our research team was keen to link numerous data sets associated with disparate spatial boundaries in an effort to include a greater range of factors correlated with population health.

Figure 4.19 illustrates the problem presented by spatial data registered to disparate, but overlapping boundaries. First, the data are compromised by assumptions of homogeneity associated with vector polygons. The population levels associated with each census tract are assumed to be homogeneously distributed over the entire spatial extent, though we know that population is typically clustered. This problem is compounded by the large spatial extent covered by each health region. Statistics on use of health services, for instance, associated with any of the four health regions are not expressed as individual characteristics, but with the whole region. The most immediate challenge in integrating diverse data sets to develop a "big picture" view of health status within the entire Vancouver urban system is to link the multiple relevant data sets to a

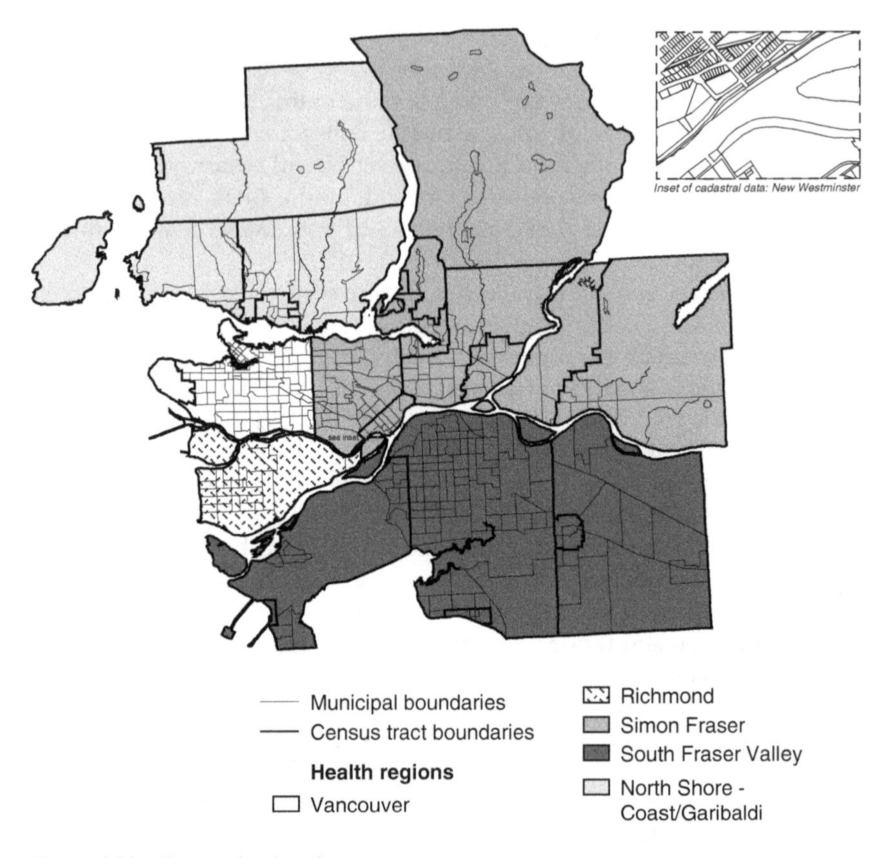

Figure 4.19 *Noncoincident boundaries.*
Numerous administrative boundaries divide the Lower Mainland region around Vancouver. These include municipal, census, and health boundaries. The map insert illustrates cadastral boundaries. Each of these boundaries is typically associated with attribute data that describes phenomena within its partial extent – but not beyond. There are a myriad other administrative divisions not shown including police boundaries, zoning areas, and school districts.

relevant spatial framework or multiple frameworks. Such a framework might contain only one or two blocks of a neighborhood when analyzing relationships between income and school performance, but may be extended to entire municipalities if availability of health services is being assessed. The spatial framework must be flexible so that areas of analysis change shape and extent depending on the questions asked. Extant boundaries associated with health regions, postal codes, or census tracts are not adaptable; their spatial definition is fixed. Using such regions as the basis for studying the health of communities implies that their administrative definition be forced upon the underlying patterns of attributes.

In order to understand how population health varies depending on the scale of the study, and the variables being used, one requires a means of allowing the attributes to express themselves as spatial patterns. This flexibility is the basis for informed policies relating, for example, to funding of childcare spaces, provision of affordable housing, or provision of school meals or remedial/enriched educational programs. In order to do this, however, a flexible spatial framework is required. One way to accomplish this is to use raster grids. David Martin, a UK-based GIS specialist has written about the use of raster surfaces to analyze socioeconomic data from national censuses. He argued (1996) that socio-economic data tend to use aggregate data, often to protect individual privacy, but a number of analyses are better suited to raster analysis. Raster frameworks also allow larger scale analysis – particularly suitable for population health studies at the neighborhood level. Even when the raster data are derived from areal data, the results are good, and ultimately more flexibility of analysis is permitted. More recently, Martin and others have suggested the advantages of surfaces for developing better pictures of population distributions (Atkinson and Tate, 2000). Moreover, converting multiple vector polygons to raster is a way of integrating multiple, disparate source data sets (Tate, 2000).

Despite a lack of precedent in using raster data for population health studies, we decided to adopt this approach in order to (i) integrate multiple datasets; and (ii) better facilitate the "data speaking for themselves." Converting the vector data to raster was, in fact, the only way to circumvent the traditional constraint faced by health researchers of working with vector polygons determined by administrative fiat. There is further justification for choosing raster over vector analysis related to the difficulties defining spatial boundaries, and in aggregating spatial units. One of the most persistent problems in GIScience is the modifiable area unit problem (MAUP). This problem occurs when spatial units such as postal codes or enumeration areas are aggregated into larger units, or when a map is redrawn at the same scale using a different spatial division. Aggregating or changing spatial units can lead to completely different results. Unemployment rates at a large scale (small number of units) may be quite high, but appear substantially lower when those units are aggregated with other data from other areas. Figure 4.20 illustrates that rates of employment seem to diverge depending on the level of aggregation associate with the data. Varying results of analysis based on aggregation of spatial units is easily used to manipulate data to convey particular messages. Clearly, when decisions about allocation of health resources are made based on spatial data, MAUP must be closely monitored.

MAUP is closely related to another problem that plagues spatial analysis: *ecological fallacy* in which aggregation or scaling introduces

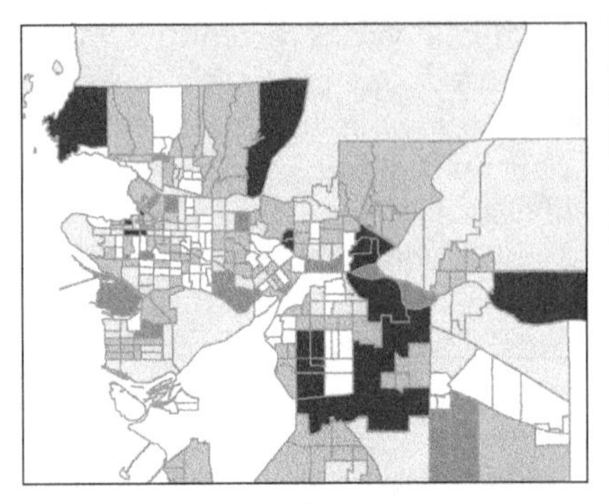

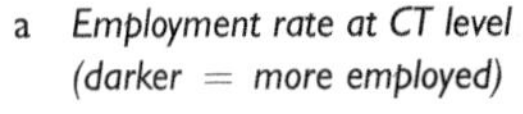

a *Employment rate at CT level (darker = more employed)*

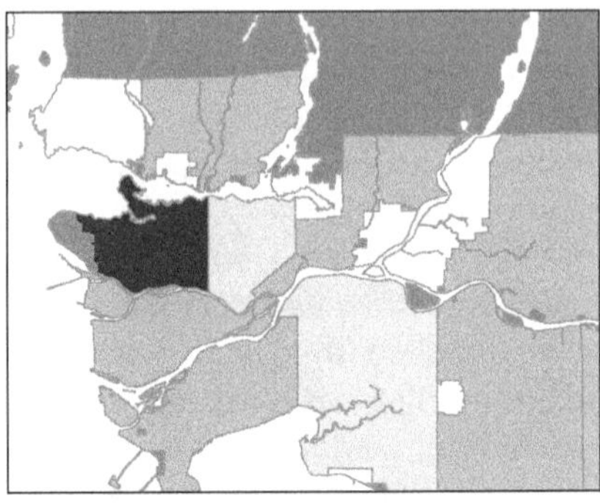

b *Employment rate at CSD level (darker = more employed)*

Figure 4.20 *Effect of aggregation on interpretation of GIS results.*
Higher employment rates in the Greater Vancouver region are indicated by darker census tract polygons in (a). Aggregation of the census tracts to the level of census subdivision yields a different interpretation of areas of high employment. The effect of aggregation on spatial variables is known as the modifiable area unit problem (MAUP).

a bias when attributing the characteristics of populations or groups to individuals. A neighborhood with a food bank within its boundaries would record many applications for assistance, but this does not necessarily reflect a high degree of homelessness or poverty in that neighborhood – just the location of a food bank. Figure 4.21 shows the percentage of the population that is known not to have completed high school. In the first map, the levels appear to be variable, and do not present immediate cause for concern. The second map uses only slightly larger spatial units as the basis for analysis, but the results are dramatically different. Indeed the large dark patches that represent high levels of noncompletion would likely cause grave concern among educators and policy makers. Different interpretations of the same data displayed using different spatial footprints illustrate a shortcoming of spatial analysis. It is difficult to determine whether characteristics of an area (such as high-incidence of TB or low rates of high-school completion) can be attributed to individual people or even specific points within the area. When flawed conclusions about individuals are inferred from characteristics of an area, this is referred to as ecological fallacy. One of the advantages of using raster grids to integrate spatial data is that both problems of MAUP and ecological fallacy are somewhat averted.

A disadvantage of using a raster based system is that the vector data must be interpolated in order to convert it to raster. If data are collected for a particular region, it is assumed that the polygon describing that region is the area for which collection was conducted. Data for mortality rates in British Columbia, for instance, are aggregated to the Health

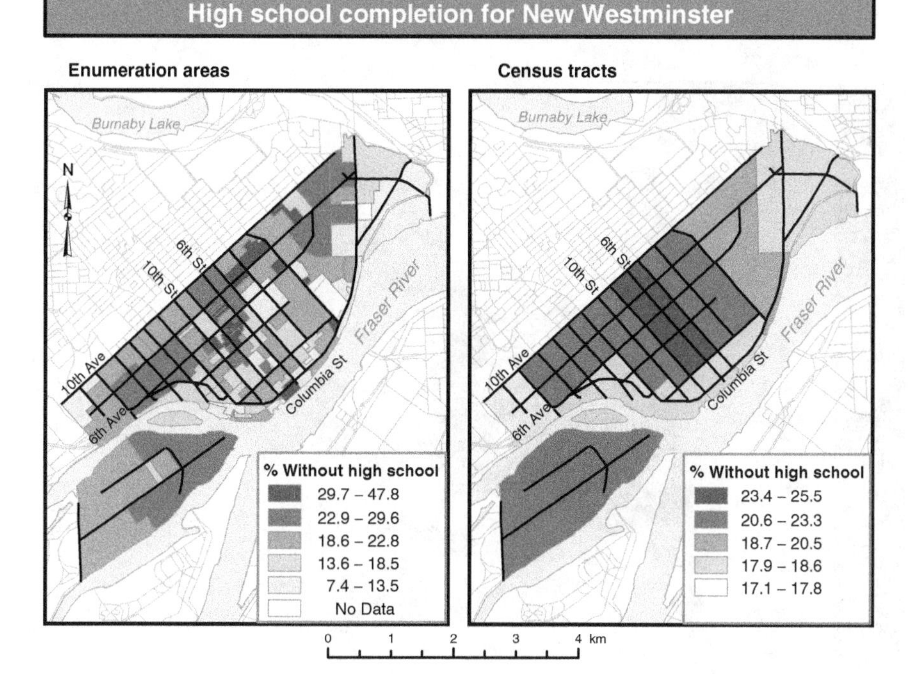

Figure 4.21 *Percentage of the population that has not completed high school shown at two levels of aggregation.*

Region level. If that coverage is converted to raster grid cells then the data must be disaggregated to create a raster coverage. This requires maintaining the assumption of homogeneity associated with those data. Each raster cell will be awarded the same overall mortality value as the health region it represents. The problem is that data collection was not done for each individual raster cell, but for the health region so it appears that we have better quality – or higher resolution – data than is the case. In converting to raster, the original data collection category of health region disappears. Figure 4.22 illustrates the mortality rate for health regions across the province of British Columbia in both vector and raster format. In this instance, the vector data have been converted to raster. The mortality data were calculated for entire health regions with the effect that when it was converted to raster grid cells the raster coverage appears to contain data at a higher level of granularity than is warranted by the collection and aggregation for health regions. The unit size of reporting no longer reflects the unit size for collection, and the seeming homogeneity of mortality rates is dissonant with true variation across the raster cells.

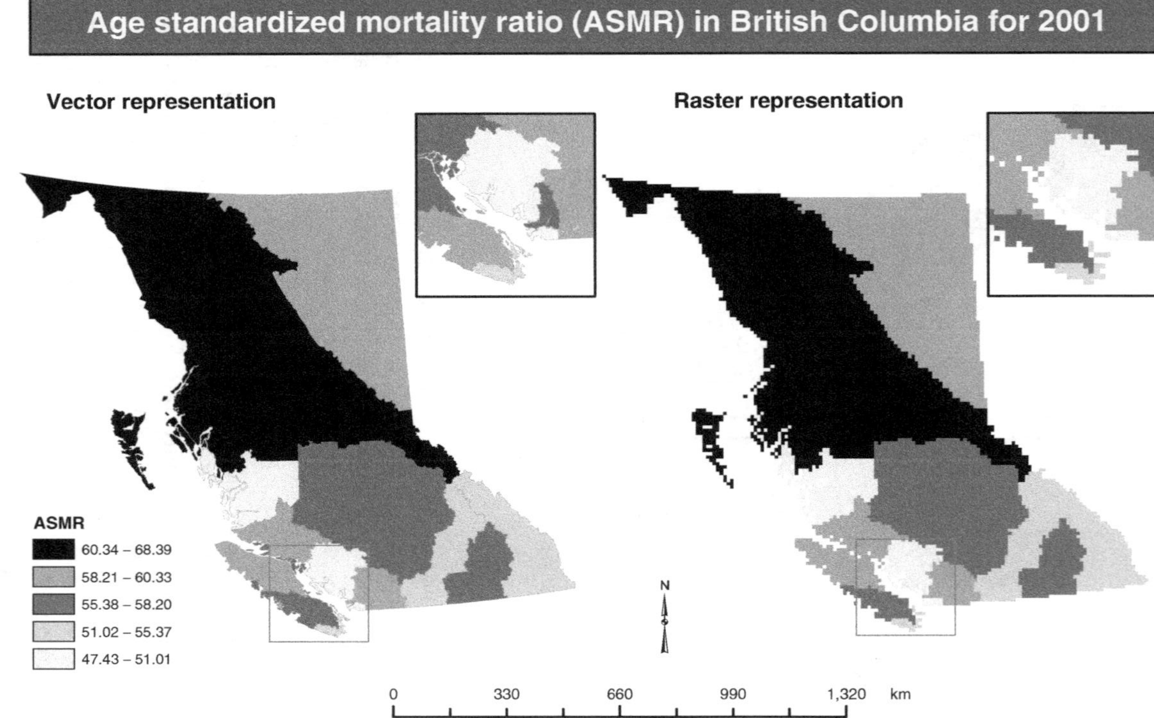

Figure 4.22 *Effect of vector to raster data conversion. When the vector data are converted to raster, the smaller raster cell size conveys the impression that the data were collected at a larger scale (smaller area) than was the case.*

A version of this problem is associated with converting population data compiled for the Canada Census to raster. In this case, both the vector polygons (such as census tracts) and the subsequent raster cells assume that data for population are homogeneously distributed across the entire polygon or cell. In actual fact, population tends to be clustered. The population health team at IHRE (Institute for Health Research and Education at SFU) was keenly aware of this limitation, and its implications for analysis of risk factors for health. They developed a *dasymetric* mapping project to correct for the assumption of homogeneity with respect to population density. Dasymetric mapping uses local knowledge to assist in areal interpolation. Familiarity with the region combined with the use of high-resolution air-borne imagery, in this case, was used to identify areas with known high or low population densities. Street network files were also used to examine housing density, and allocate population based on that information. Likewise, liaisons were established with each of the health regions to negotiate access to vital statistics about births and deaths based on individuals linked to postal codes rather than aggregated to the regional level. The interpolation from vector to raster was not an algorithmic operation, but a mix of human intervention and automated conversion of data.

MCE and Analysis

With these integrated data our current goal is to analyze the distribution of health status in the GVRD using a variety of indices and MCE. In this case, we are chiefly concerned with nonmedical determinants of health and community/health system characteristics, and of health system performance linked to the provision and availability of health services in the province.

Motivating questions include:

1. do quality of life indicators vary between neighborhoods and municipalities in the GVRD?
2. how can we link quality of daily life indicators to observed spatial differences in health status and use of health services?

Traditionally, neighborhoods with compromised health profiles have been identified using indices that are applied to fixed vector spatial frameworks such as census tracts or postal codes. Borders for polygons of this type are usually determined by administrative fiat. Because of this they are seldom ideal units of analysis for health data as communities identified with robust or marginal health seldom restrict themselves to administrative jurisdictions. Raster data formats are a better basis for identifying self-organizing constellations of factors affecting community health (see above). One of the problems the project faced was that most health indices have been developed to work with vector data.

One of the better known indices is the *Jarman 8* developed to measure underprivilege. It was developed in the UK to identify underprivileged areas for the purposes of health care planning. The eight variables in the United Kingdom index were selected on the basis of a national survey of general practitioners regarding factors affecting their workloads. Each variable in the index is weighted according to the average degree of importance given to it by the surveyed general practitioners. The model adopted by the IHRE at SFU is based on the Australian version of the index, mainly due to the similarities between the 1986 Australian and 1996 Canadian census variables. The eight variables calculated for use in the Australian index are listed in Table 4.2 (with their weightings in parentheses):

The Jarman 8 index was modified somewhat to render it suitable for the Canadian context. We used 7 factors from the Jarman 8 Index so that results from the two methods could be compared. For the first factor "Elderly alone," the age was changed to "65 years and over" instead of "60 years and over"as this reflects categories used in the Canada Census. The Jarman factor we didn't use in the MCE is "Ethnic minority" because the definition is ambiguous in the Canadian context. Canada has such a vast ethnic diversity that it is difficult to determine which category

Table 4.2 *The eight variables (and their relative weights) used to calculate the Jarman 8 index of underprivilege used for health care planning and resources allocation*

Persons aged 60 years and over living in lone person households as a percentage of all persons aged 60 years and over in private dwellings **(6.62)**	Children under 5 as a percentage of total persons counted **(4.62)**	Single parent families as a percentage of all families **(3.01)**	Laborers and related workers as a percentage of all employed persons **(3.74)**
Persons aged 15 and over who were unemployed as a percentage of the total labor force **(3.34)**	Average number of persons per bedroom in occupied private dwellings **(2.88)**	Persons aged 1 and over counted at home with a usual residence one year before the census different from present usual residence as a percentage of total persons counted at home (minus persons aged under 1) **(2.68)**	Total persons born overseas minus persons born in Other Oceania, United Kingdom and Ireland, and Other America as a percentage of total persons counted **(2.50)**

constitutes a minority, and the minority designation is unstable over space.

Figure 4.23 illustrates the results of the index for New Westminster, a municipality in the GVRD. The chief drawback of the index is that it takes the census enumeration area (EA) boundaries as *a priori* spatial definitions. The assumption is that the EA boundaries developed to expedite

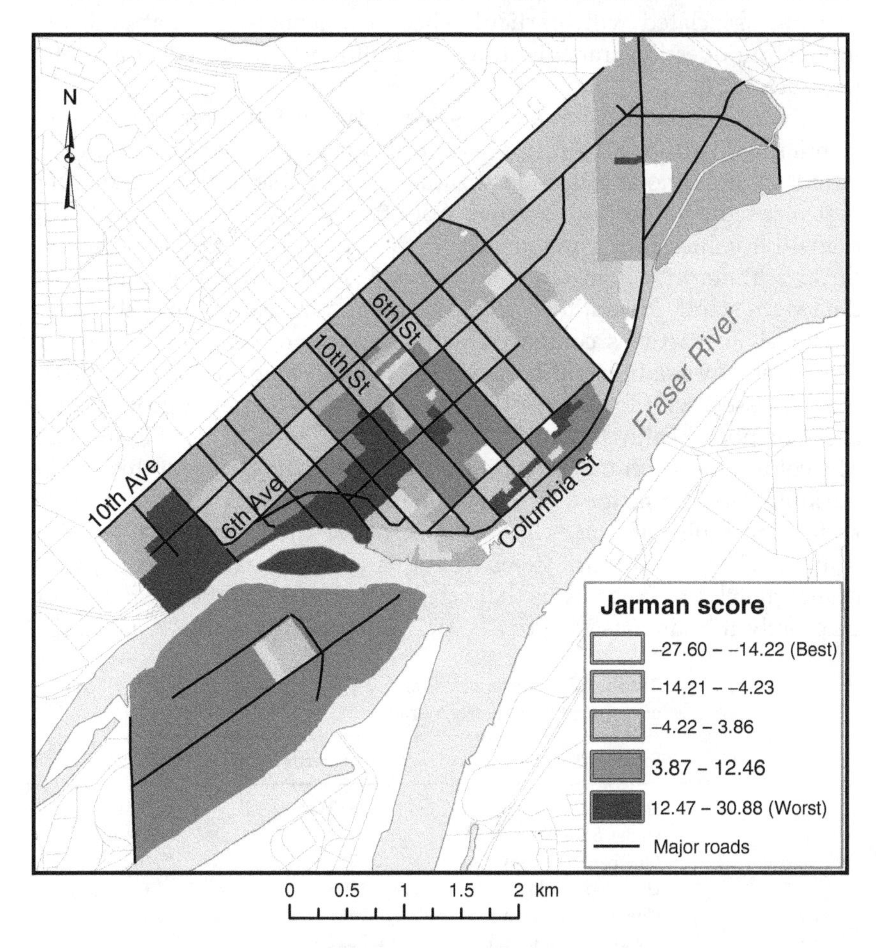

Figure 4.23 *Results of the Jarman index calculated for New Westminster, a suburb of Vancouver. The index was developed to measure underprivilege associated with a particular spatial area. The darker colours indicate areas of greater underprivilege according to the parameters of the index.*

the collection of socioeconomic data constitute a rational basis for organizing neighborhoods based on potential health outcomes.

If we return to our motivating questions, it is clear that it makes more sense to let the data define the neighborhoods than rely on those defined by administrative fiat. The advantage of having converted our data to raster is that there is less ontological commitment to boundaries associated with raster data (see Chapter 2). While vector-defined neighborhoods used by health researchers quickly become naturalized, we know that their very existence is contingent on factors unrelated to health. From a GIS perspective, one way of allowing the data to "speak" is to use raster queries to isolate clusters of like variables that are associated with particular health outcomes. To that end, our motivating questions must be translated into factors and constraints for use with MCE.

A multi-criteria decision analysis requires that the criterion scores be standardized so that the factors are in the same scale and can be compared. Two methods of standardization are suitable in this context: the first maps the values into a range from 0 to 255 while the second type maps the values into a range from 0 to 1. Both methods use a simple linear scaling. In this instance, the second method was used. Once the data were standardized, they were weighted as illustrated in Table 4.3.

The MCE map was compared with the Jarman Score map to find if there was any relationship between them. Both maps are broken down into 5 classes, with the areas of good health shown in light grey and the areas of poor health shown in dark grey as illustrated in Figure 4.24. The color patterns in the two maps appear quite different but this can be attributed to two factors. First each map uses a different measurement scale. The range of values is different hence they can't be compared directly. The second factor stems from this scale mismatch: the categories do not match up. This means that category [191–209] in the MCE will not be exactly the same as category [−4.22–5.44] in the Jarman. A reclassifi-

Table 4.3 *Standardized variables for the MCE analysis and their relative weights*

Factors	*Normalized weights*
Elderly alone (la65)	0.3933
Under 5 (und5)	0.1650
One parent (onep)	0.0776
Unskilled (unsk)	0.1106
Unemployed (unem)	0.0957
Overcrowded (ovcr)	0.0822
Changed address (chad)	0.0755
	1.0000

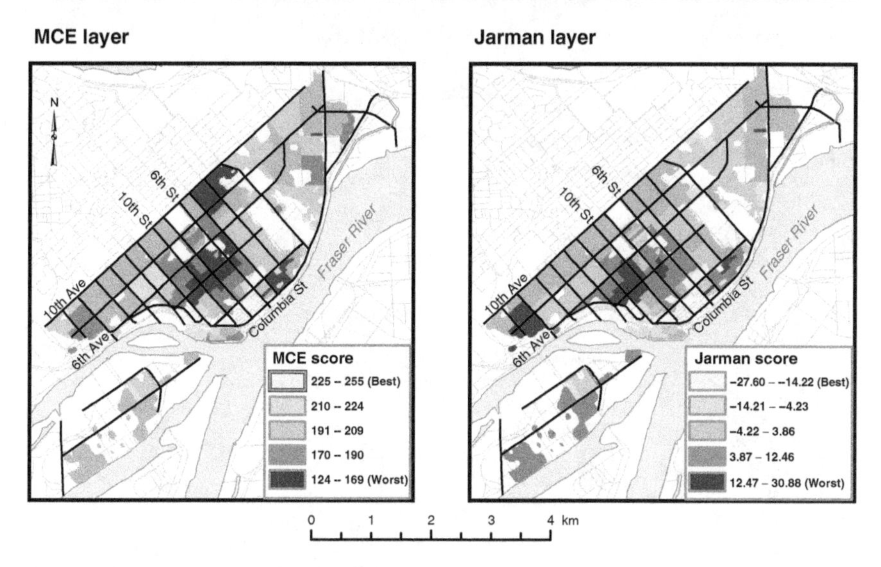

Figure 4.24 *A comparison of multi-criteria evaluation (MCE) with the Jarman index. They produce comparable though differentiated estimates of area of underprivilege. The advantage of MCE is that it allows factors to be adjusted to reflect different perspectives and agenda. It uses a raster data structure which permits the use of variable cell sizes as the basis of analysis.*

cation of the category values does not help since the distribution of values within the categories varies between the two methods of creating the maps. To solve this problem it would be necessary to standardize the scores from the two methods to the same scale. However, it is difficult to justify doing this since the two methods are quite different. Both maps are legitimate representations, but not comparable. The advantage of the MCE generated image is that its categories emerge from constellations of attributes within the data rather than relying on those developed for a very different purpose: the resulting neighborhoods are self-generating and therefore more useful to health researchers.

Some of the inherent challenges in interpreting GIS results are highlighted by this example. The most salient lesson is that neither the Jarman index nor the MCE analysis can substitute entirely for local, on-the-ground knowledge. It is well known, for instance, that areas of high urban density are associated with higher risk for poor health outcomes. Given this common epidemiological supposition, it follows that areas of high-density should be associated with poor health outcomes. Figure 4.25 depicts two maps resulting from analysis of (i) density and (ii) "at-risk" for poor health based on the Jarman variables for New Westminster.

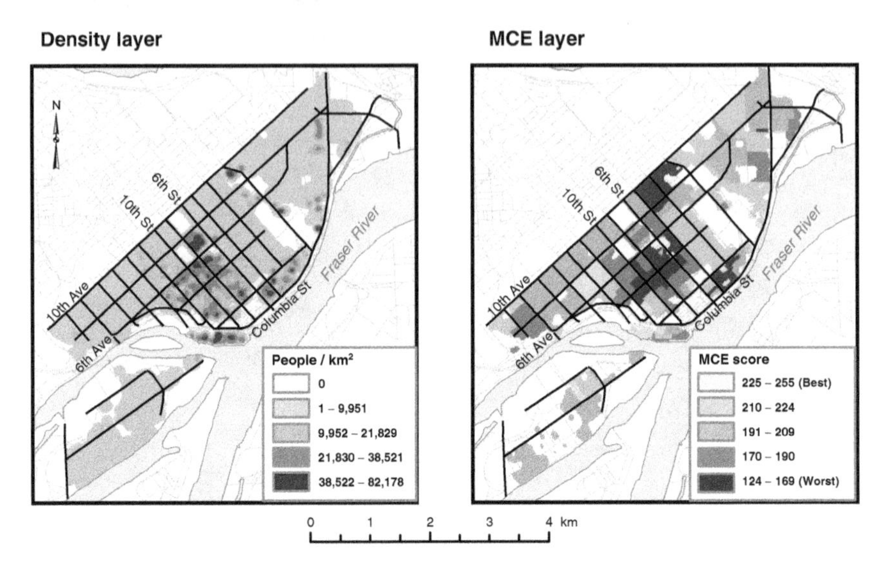

Figure 4.25 *The population density and MCE analysis of underprivilege show a correlation between high population density and health risks associated with underprivilege. This relationship does not hold, however, for all areas of the Greater Vancouver region. In Vancouver's downtown, for instance, density is linked to low health risks.*

Note that high values in both maps coincide geographically. Also, note that land use values have been used to discern and eliminate the land not associated with housing from the analysis. This permits the population density values to be redistributed to better reflect where people live. From a casual survey of the results, one might conclude that densification of urban areas does indeed predispose residents to greater health risks. The near perfect match between these two values certainly supports this conclusion, but maps of the same indicators in the downtown area of Vancouver do not corroborate this coincidence. There, extreme densification is correlated with relatively robust population health especially if resident age is accounted for. The results in Figure 4.25 are related to the income and real estate values associated with a particular form of densification. The downtown area of Vancouver demonstrates different relationships between income and real estate values resulting in a negative correlation between density and health risks. The discussion on visualization and exploratory data analysis (above) points to the need to combine visualization with confirmatory testing of relationships. This example also points, however, to the need for continual critical examination of GIS as a form of calculation that is performed in the interests of certain groups based on particular assumptions.

Calculation and the Rationalities of GIS

This chapter has focused somewhat uncritically on the calculating technologies inherent in GIS. Many of these are extremely useful in describing and modeling the complex interplay between diverse phenomena. The examples presented here are illustrative of the utility of such analysis in the context of environmental management, resource allocation, and maintenance of public order. They are functional, however, only in a particular sociopolitical context. In order to step back from the powerful combination of quantitative analysis and visualization that constitute GIS, it is useful to ask whether the Hawaiian Islanders that Captain George Vancouver encountered in the eighteenth century would be persuaded by GIS generated maps of land suitable for pineapple cultivation. Likewise, are present day arguments on the part of the United State federal government researchers that Nevada subsurface rock strata are ideal for storing spent nuclear waste convincing to First Nations peoples and other local residents? Context and frames of reference contribute to the legitimacy of any system of representation. Canadian literary critic Northrop Frye (1982) wrote that in church-dominated thirteenth century Europe, there were nineteen classifications of angels, and none of geological strata! What and how social and natural phenomena are represented depends on the priorities and vision of a society and its governing structures.

Mark Poster (1996) has written about the increasing tendency of governments to view citizens as constellations of statistics (e.g., your age, Social Insurance Number, income, profession, tax rate etc.). Poster calls these collections "digital citizens." Your description as a digital citizen is likely hollow when compared to the "flesh and blood" you, but it nevertheless substitutes for you when you apply for a bank loan, an entrance scholarship, a passport, and in numerous other modern situations. Governments are not alone in the role of data collector. Marketers are masters of digital data collection and integration of multiple sources is made easy by the ubiquitous use of Social Insurance Numbers (Social Security in the US and UK) for identification. Each time a person uses a "loyalty card" or supermarket savings card, his or her purchases are recorded and added to a growing folio that describes their buying patterns (spatial and attribute). The trend toward Smartlabels exacerbates this trend. Smartlabels are new merchandise tags that contain inventory information so that large supermarkets can better control supplies. Smartlabels retain their intelligence, however, after they pass through the till. Specific products are linked to the customers that buy them based on customer data associated with debit and credit cards (as well as loyalty cards). Products bought with Smartlabels can be traced to individual households – a spatial form of consumer intelligence gathering. Early opposition to Smartlabels has resulted in the development of a "kill" command which allows the customer to turn-off the Smartlabel at the till – with the attendant threat of poor inventory control at home. In the post 9/11 era, the trend toward data collection has increased and GIS plays a vital role in this form of surveillance. It is simplistic, however, to cast aspersions on the technologies implicated in these data warehouses. Technologies are the outcome of social processes.

Governments and businesses operate within specific "rationalities." Jeremy Crampton (2002) explains that practices associated with a particular time and place are reflections of larger ways of thinking – rationalities. These can be situated historically as Michel Foucault (1979) superbly demonstrated, and they are dynamic. The present era which Crampton designates as one of "carto-security" is, for instance, embedded in a paradigm of an enemy among us yet difficult to detect. The spatial surveillance technologies being developed and deployed reflect this political issue. Likewise, development of statistical techniques in the nineteenth century were closely related to the political issues of the time. Technologies and social imperatives have a way of dovetailing.

Michael Curry (1997), a well-known philosopher of geography confirms that GIS clearly supports a trend toward a more surveillant society. Adding a spatial component to databases allows users to locate and describe *individuals* rather than demographic clusters. The result has been a trend toward "rooftop" marketing in which the consumer characteristics of individual households are documented. GIS has enhanced the precision of marketing information. In the new era of carto-security, scanning passports at borders, finger-printing members of specified ethnic groups, and visual and electronic bio-identification measures are contributing to detailed accumulations of digital data on citizens by national governments.

The chairman and chief executive of Sun Microsystems declared several years ago: "You already have zero privacy – get over it" (*New York Times*, March 3, 1999). This soundbite, expresses an inevitability about a loss of privacy, but it is illustrative of a particular social rationality that does not apply everywhere. The European Union passed in 1995 and again in 1998 *Directives on the Protection of Personal Data* which require the consent of individuals for the sales of personal data. In order to sell data that are safeguarded in Europe, assurances must be obtained that the data will be similarly protected in the destination country. These directives are pointedly aimed at the United States but also reflect traditions of privacy ensconced in Article 8 of *European Convention for the Protection of Human Rights and the Fundamental Freedoms* in which the right to privacy at home, in correspondence and in family life are safeguarded. Through privacy regulation, the European Union and to a lesser extent Eastern Europe has defined itself as a community prepared to defend privacy in the cyber-world.

Michael Curry (1997; 1998) offers another partial clarification to the mystery of why some citizens are better protected from surveillance than others, pointing out that ethics only make sense in particular cultural contexts. Europeans operate within a rationality that stresses protection from government and commercial surveillance. Citizens of the United States have recently suffered an attack and are intensely aware of on-going security threats. Moreover, their economy is based on a philosophical preference for *laissez faire* economics in which data are interpreted as a commodity. These differences are, however, not constant. They reflect changing social and political rationalities. As Michael Curry has noted (1997), the line between the public and private has shifted throughout history. People can and will adjust their behavior to take into account the level of surveillance which is implicit in their lives.

GIS and its implementation are constituted by prevailing rationalities. Its calculating methodologies are servants of social goals. The breadth of GIS usage including PPGIS (Public Participation GIS) environmental management, health studies, military navigation, and rooftop marketing are indicators that we live in diverse societies with multiple and conflicting rationalities.

5

Where Do I Go From Here? GIS Training and Research

The previous chapters have consolidated a vision of GIS that is rooted in intellectual practices, populated by data, and powered by mathematical analyses. This final chapter provides readers with further insight into the relationship between GISystems and GIScience as a means of understanding how they might participate in this burgeoning branch of geography. The chapter explains the training necessary to use the software that constitutes GIS, and emphasizes the skills required to use GIS accurately and persuasively. This discussion is balanced by the need to think outside the icons and command buttons that fill the screens of popular packages. On-going research in GIScience provides the basis for addressing many of the limitations of present-day software in terms of representing geographical phenomena and the myriad constituencies that are affected by them. To that end, two distinct but influential areas of GIScience research are introduced: ontology research and feminism and GIS. Each of these pursuits is based on very different rationalities and motivations, but they share the potential to extend the purview and practice of GIS. Finally, the interdependencies between GISystems and GIScience are stressed; both are essential for the continued use and improvement of this substantive area of geography.

People and Research Versus Software Training

Even a cursory reading of this book thus far informs the reader that GISystems and GIScience are closely related – but not equivalent. In this final chapter, that relationship will be more fully explored with the goal of emphasizing their mutual dependence. In the simplest of terms, GISystems are the software and hardware, and GIScience the theory and intellectual assumptions that underlie them. This dualism, despite its oversimplification equates GIScience with research and academic training, and GISystems with software and particular vendor implementations. GIS instructors can testify to the vast divide between the abstract expression of GIS principles and their execution. Experienced users are of considerable value to many businesses and organizations while a GIScientist might seem abstract and peripheral. Both are essential to the continued use and development of GIS.

This gap between the principles of spatial analysis and its execution is expressed in the following interchange. Stan Openshaw, a preeminent spatial analyst, was optimistic about the potential of GIS for numerous research areas. According to Openshaw, a geographer of the impending new order in which geography is dominant may be able to:

> analyze river networks on Mars on Monday, study cancer in Bristol on Tuesday, map the underclass of London on Wednesday, analyze groundwater flow in the Amazon basin on Thursday, and end the week by modeling retail shoppers in Los Angeles in Friday. What of it? Indeed this is only the beginning. (Openshaw, 1991, 624)

Stacy Warren, a geography professor specializing in GIScience at Eastern Washington University, responded to this ambitious plan by emphasizing that she would still be labeling Tuesday's cancer polygons on Friday. Warren's comment summarized the distance between the ambition of spatial analysis and its implementation. Many papers in the GIScience literature present theoretical solutions to difficult problems such as interoperability or including time as a dimension of spatial entities. Implementation of those solutions is the sticking point.

Procrastinating students in introductory GIS courses are often horrified the night before their first project is due. At that point, the difficulty of working with GIS software becomes apparent. GIS analysis is completely different than writing an essay. The latter, however poorly written, can be done quickly; we are all familiar with conventions associated with speaking and writing. A terribly misguided sentence will still appear on the page of the word processor. In GIS, however, folly is self-aborting. If you don't build topology tables in ArcInfo®, for instance, the analysis will not follow. Each program is structured on precise user requirements that must be fulfilled. This is true of all information systems. When writing a computer program, a ":" is interpreted completely differently than a ";." Careful, painstaking review of the order of instructions is often required to discover the reason for the dreaded message "run-time error." A long evening can pass searching for possible errors such as the letter "o" mistakenly substituted for a zero. As GIS software becomes higher-level with icons and user-friendly menus replacing line driven codes, such impasses become easier to avoid. The need for experienced GIS users, however, is not expected to dwindle.

GIS users and researchers are both valuable and mutually indispensable, but emerge from different venues. Students interested in learning how to execute spatial analysis in specific software environments are encouraged to learn GIS in the systems interpretation. There are multiple avenues for doing so. Some users learn on the job using off-the-shelf software, while the vast majority attend a technical school or college where they are trained in the organization of spatial data, databases, and in the execution of common analysis operations. Many of the most proficient users in business and government received minimal academic training in GIS, and simply learn by doing.

Despite the important niche occupied by proficient operators, GIScience is integral to the reliability of results. The most seamless implementation of data and analysis in a GISystem does not substitute for the ability to justify choices of resolution, data integration criteria, data model, type of analysis, and cartographic representation. In the absence of these pillars of GIScience, the results may be superficially persuasive but remain poor substitutes or supplements for other types of geographical analysis. It is a delicate balance between developing the hard-won and valuable skills of a GIS operator, and pursuing the more abstract but equally precious intellectual proficiency that constitutes GIScience.

The elucidation and extension of the intellectual premises of the software are pursued under the rubric of GIScience. Universities tend to stress the underpinnings of GIS in a more formal and abstract manner with greater emphasis on abstract principles and the potential of GIS for understanding complex phenomena. This training is not necessarily commensurate with the degree of practical experience required to be hired into a busy operational environment, but it does ensure an abiding appreciation for the potential of GIS to elucidate complex spatial relations, aid in decision-making, and explicate patterns through visualization. These are the best known tenets of GIScience, but it is a large tent and many forms of investigation have taken umbrage under the rubric.

The expansion in areas of scholarship associated with GIScience is multifaceted but can be approximated through a content analysis of its most representative journals. To that end, the categories shown in Table 5.1 were chosen for the content analysis with the caveat that all categories are subjective, situated perspectives.

Each of these categories is fraught with overlap as well as some ambiguity. Their selection does, however, represent common categories of GIS research. Four representative journals were chosen for the content

Table 5.1. *Categories for content analysis of GIS journals, 1995–2001*

Applications of GIS	*Spatial analysis and modeling*	*Data*	*Cartography and visualization*
GIS and society	Ontology and epistemology	Cognitive/spatial reasoning	Algorithms

analysis. Each has been in publication for at least seven years, and is well known in the GIS and geography communities. Stock journals such as *Environment and Planning A* and *B* were excluded despite having many papers dealing with GIS as they are not *primarily* GIS journals. In the end, 566 papers were analyzed from four journals: *International Journal of Geographical Information Science (IJGIS), Transactions in GIS, Cartography and Geographic Information Science (CAGIS),* and *Cartographica.* The long and close relationship between cartography and GIS is emphasized by the inclusion of *CAGIS* and *Cartographica.* The distinctive mark of each included journal is its primary commitment to GIS and cartography.

The results of the analysis are illustrated in Figure 5.1. As this graph indicates, the vast majority of research in GIScience is predictably focused on spatial data, analysis, algorithms, and cartography. Equally telling, however, is the limited number of papers that dealt

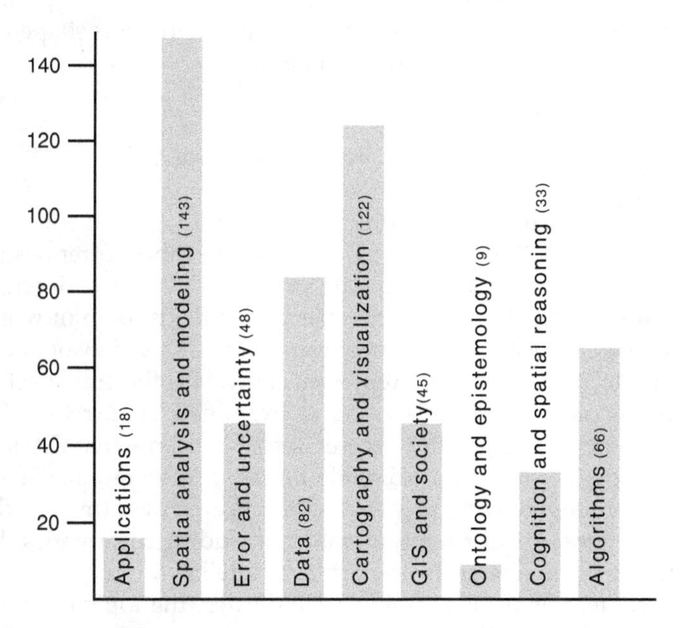

Figure 5.1 *Content analysis of GIS journals from 1995–2001 (based on 566 papers).*

with applications, flippantly referred to as "How I used GIS to count reindeer in the Arctic" papers. Despite the importance of such papers in addressing social, political and environmental issues like environmental justice or land reform, they are not considered central to GIScience. Instead GIS implementation papers offer instruction on the range and possibilities of existing software – just as they point to its shortcomings. Moreover, these types of papers are seldom critical of existing GIS software or its intellectual assumptions. Rather they implement the systems to solve a geographical problem. They are distinguished from GIScience by their emphasis on spatial problem solving rather than extending its premises and intellectual bases.

The shorter bars on the histogram are of particular interest in this context because they represent growing areas of research, areas that firmly establish GIScience as interdisciplinary, and of relevance to all geographers. The following sections explore two areas of current research in GIS that are particularly of interest to human geographers: ontologies and epistemologies; and feminism and GIS. These two areas of investigation may appear superficially unrelated, but both offer means for extending GIS to include greater possibilities for representation. Both strive to incorporate multiple points of view – an axiom of feminist politics. Ontologies and epistemologies are relevant to the GIS users because they identify "the object of inquiry." Feminist analysis of GIS illustrates the ways in which knowledge production is always partial. It is both selective – not omniscient – and subjective in that it is shaped by the social, political, and economic environment.

Ontology and Epistemology Revisited

In Chapter 2, the related philosophical axes of ontology and epistemology were introduced in the context of GIS. Their relevance to representation and the role that data models and data structures play were described. In a word, GIS divides the world into objects and fields, both of which can be represented by raster or vector data structures. This view of ontologies as data models does not fully represent, however, the full spectrum of ways that ontologies are understood or examined in current GIScience research. Rather it is a practical expression of concerns that GIS scholars raise about the parameters of digital spatial representation. These concerns have proliferated into an active research area over the past decade, one with implications for many domains including informatics, human geography, cognitive science, and artificial intelligence.

As the content analysis in Table 5.1 indicates, the topic of ontologies did not dominate academic research papers from 1995 to 2001. The corresponding histogram is, however, somewhat misleading for two reasons. First the bulk of papers dealing with ontologies were published in a special issue of *IJGIS* in 2000. As well, mention of ontologies in other papers that deal more specifically with data or spatial analysis lead to an underestimation of its import. Second, and most significantly, the content analysis fails to include subsequent papers published in 2002 that deal explicitly with ontologies. Most significantly, the second biannual GIScience conference held in Boulder, Colorado included over a dozen paper presentations that unequivocally acknowledged the importance of this area of research, and offered insights into its clarification. This conference is significant because it (i) focuses only on GIScience issues and (ii) brings together GIScience researchers from all around the world.

Ontology as an emerging niche of GIScience is understood and interpreted differently by different scholars. The range of interpretation of "ontology" varies widely, and includes but is not limited to: (i) a proxy for data models and debates about the limits of their power to represent spatial phenomena (described in Chapter 2); (ii) description of cognitive and perceptual impressions of space and spatial entities; and (iii) as a surrogate for classification systems and taxonomies. These noncongruent ways of construing ontology are further complicated by the distinction in information science between philosophical ontologies and formal computational ontologies.

Philosophical interpretations of ontology draw on a long history of western philosophy which stresses that, in the absence of interpretation or mediation through the senses, there is an *essence of being* associated with every entity. Formal computational ontologies are based on a fixed delineation of entities within a knowledge domain, where the definition of each entity and its relationship to every other entity is proscribed. In the digital world, these ontologies translate into invented universes in which spatial entities are crisply defined, and their range of behavior as subjects is similarly circumscribed. For example, an ontology might simply define a bridge, a road, a lake, and a river. In this case, the relationships – both topological and functional – between each of these entities would also be defined. The lake, for instance, may drain into the river. The road may run under the bridge. This is a very different notion of ontology than is commonly employed in the humanities.

Prioritizing these different understandings of ontology is impossible, but recognizing that the word means different things to different people is important. The difficulty of pinpointing meaning in this context is expressed by the tongue-in-cheek title GIScientist Gary Hunter (2002) also used for a recent editorial in *Transactions in GIS:* "Understanding Semantics and Ontologies: They're Quite Simple Really – If You Know What I Mean!" Indeed, variable interpretation of "ontology" speaks to the heart of the problem of semantic integration that plagues all information science – not just GIS.

Some academics argue that ontology is a new term for what were formerly data models, formal specifications and semantics. Philosophers, however, might argue that this is a simplistic interpretation of a complex subject area. There is, however, a justifiable pragmatism associated with using data models to stand in for ontologies. In GIS, representation must always be filtered by a data model. Representational ontologies are as limited as the data models that serve them. Werner Kuhn (2001) sees the ontology debate today as an extension of historic discussions about feature attributes or characteristics. In this version of ontologies, the debate remains framed by the scope of digital data models and structures. He rejects this equivalence between data models and structures, however, by arguing that ontologies tend to be designated as nouns. Lake, pond, building river, bridge, community, and region are all instances of spatial attributes that are considered part of an ontology in particular domains. Kuhn points out, however, that most spatial entities are associated with *processes* not just location or other attributes. He argues that by including process information with ontologies, their meaning will be clearer for the purposes of data integration between domains. In effect, including process information will allow users to better assess what a semantic term was intended to mean.

Kuhn (among others) incorporates the idea of *affordances* to illustrate that different ontologies offer different types of logical operations and representation. For instance, a wildlife habitat in British Columbia *affords* or supports grizzly bears and elk. A road affords automobile traffic, or, in other instances, affords only bicycle traffic. When semantic terms are linked to affordances, they offer a richer description, and also provide a basis for future semantic integration. Kuhn suggests that domain information such as affordances should be included in data models based on the logical capabilities of different systems. This approach is attractive as it works within existing frameworks to better represent the data that populate them. A new field could be added to existing databases to describe the affordances of spatial entities. The field would imply context associated with the field as well as appropriate use. The problem with this method is that it is social not technical: it does not account for the *ad hoc* manner in which most data are compiled. And by focusing on singular processes, it fails to account for unique and diverse interpretations of spatial phenomena – epistemology in effect.

Martin Raubal (2001) extends Kuhn's model by arguing that affordances are shaped by epistemology. Affordances are dimensionalized in this context by paying attention to ways that culture and social context affect them. Raubal uses the example of a mother and her preschool son navigating an airport. For the mother, the ticket counter *affords* transactions with the airline agent, as well as a resting place for her purse. For her son, the counter is a vertical barrier to participating in transactions. Raubal's research in this instance is directed toward developing automated agents (individualized software tools) for assisting persons navigating large public spaces such as airports or libraries. He draws on cognitive and ecological research to develop ontologies and affordances for the objects, people, surfaces and navigational routes that comprise such environments. Ontologies in this instance are specific to navigational tasks, and affordances reflect epistemologies of users in *particular situations.* Both ontologies and epistemologies are developed to inform the software agents with the goal of assisting users in navigating complicated public spaces that are unfamiliar.

Other geographers argue that ontology research should not be premised on how computers handle data and the limitations thereof. Instead, we should ask how people perceive and express spatial entities, and make GISystems sensitive to the categories that people already use. Barry Smith and David Mark first argued in 1998 that ontologies must account for human practices, both administrative and cognitive. In 2001, they went on to suggest that naïve (i.e., untrained) geographical perspectives vary from that of experts, suggesting that ontologies should be based on the categories used by potential users rather than information scientists or domain experts. Smith and Mark conducted a study that illustrated that nonexperts conceptualize geospatial phenomena using common primary spatial categories. When asked to list geographic features or objects, for example, most subjects supplied examples from physical geography such as mountain or lake. When asked to list "things that could be portrayed on a map" respondents were more likely to respond with roads, cities, as well as some physical geography features. In the most general sense, their research suggests that external conceptualization of a domain must be considered when defining objects/ontologies that will be used by a broader population. The shortcoming of this research is that it assumes that test subjects at Universities in Uganda will respond the same way as a sample of American University students. The authors do attempt to justify this assumption based on a level of universality in cognitive perspectives. Yet, the fact remains that people and domains have divergent interpretations of how to define and represent spatial phenomena.

Recognition that domains vary in their understanding and representation of similar phenomena motivated Frederico Fonesca, Max Egenhofer, Peggy Agouris, and Gilberto Camara (2002) to explore a method of accommodating divergent ontologies in a GIS: one that does not rely on shared data models nor a singular conceptualization of geographical reality as the basis for integrating multiple ontologies. In their proposed "Ontology-driven GIS", ontologies refer to a classification system or taxonomy, a dictionary or thesaurus. The system has two primary components: knowledge generation and knowledge use. *Generation* is based on the specification of ontologies using an ontology editor. Developers of the system would be responsible for knowledge generation which is a form of extended metadata that includes information on the classification system and other relevant information that is usually missing from data. The *use* phase allows users to manipulate the formalized objects and classes to fulfill information requests for the purpose of data integration or data seeking. One of the attractions of the proposed ontology-driven GIS is that data can be integrated regardless of their native data model. Both object and field data can be described and contextualized using the ontology editor. Moreover, the system takes into account multiple cognitive universes in which people differently perceive the physical universe. The authors regard ontologies as a way of conceptually surpassing the map metaphor that has dominated GIS. The proposed system would permit semantic interoperability based on: (i) understanding semantics of the information source; and (ii) use of automated mediators to return information requests based on ontology information ensconced in the editor.

Ontology editors that work with any data model follow an emerging trend toward surpassing the binary or dualism dictated by distinctions between objects and field. Tom Cova and Michael Goodchild (2002) developed a method for linking object and fields to create object fields that derive objects from a raster data model. These object fields can be dynamic, and can also have fuzzy boundaries. They are described computationally with a simple and elegant system of formalization. Rather than consider the two prevalent data models to be entirely representative of geographical reality, they are consigned to pragmatic, conceptual models which permit the users to develop temporary and functional objects to describe certain phenomena. Object fields take advantage of the fact that people and language are better at describing objects than fields; objects have more inherent meaning than correspondent clusters of attributes in a raster data structure. The authors offer the example of objects that might result from the spill of hazardous material as illustrated in Figure 5.2. The extent of a given spill would vary depending on the surface and elevation. The object field is calculated independently for every raster cell in the coverage from which the spill could originate. In this case, three possible object fields are illustrated.

Object fields allow GIS users to understand how phenomena vary across a study area. They also permit informed users to update the defining parameters of the object field if they observe that it is does not

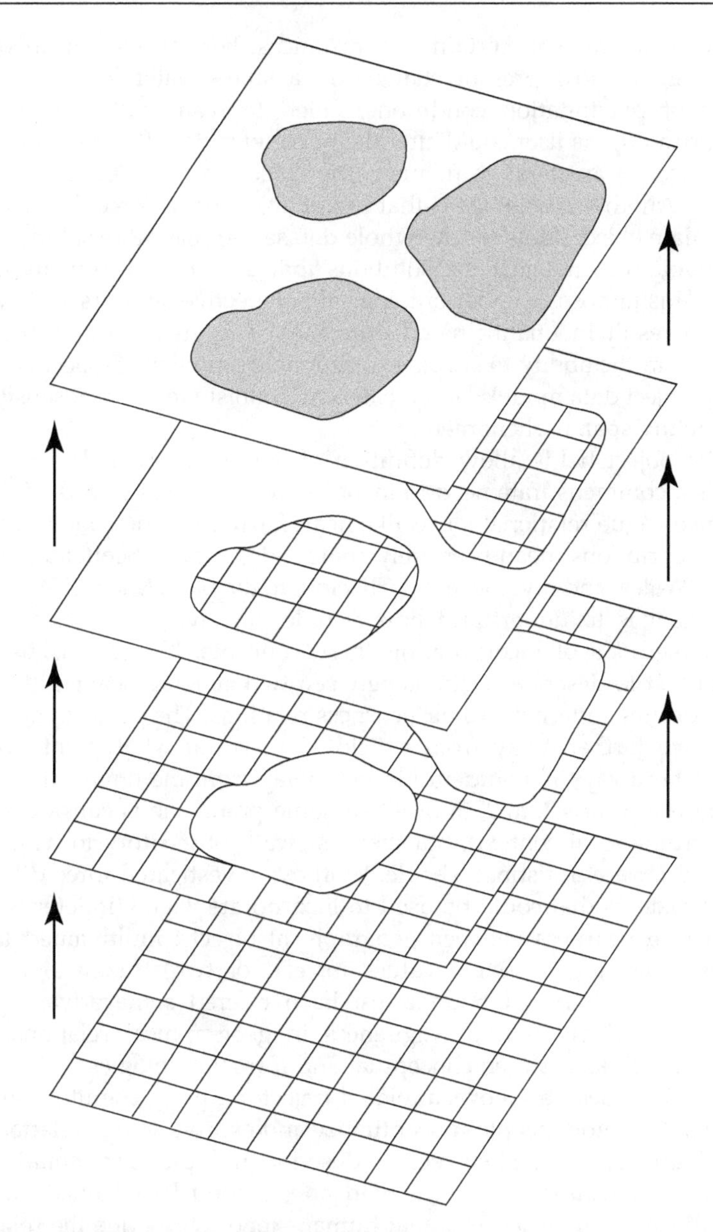

Figure 5.2 *A conceptual outline of object fields as described by Cova and Goodchild (2002). Here a base raster coverage is overlaid with other significant attributes to yield three object fields that describe the extent of hypothetical spill of a hazardous material.*

behave realistically in certain circumstances. For instance, if an object field designed to represent changes in a lake's water levels, given a variety of precipitation conditions, failed to realistically account for spring run-off, the user could alter the parameters to reflect her empirical experience. In an iterative manner, the object field can be made more realistic. Another advantage is that use of object fields force the program to calculate object fields for the whole dataset, so that more solutions are considered (i.e., not just the solutions that user has already hypothesized). This imposes a wider ontological perspective on users or viewers, and can result in unanticipated outcomes. The greater implication is, however, in the ability to surpass traditional restrictions associated with field or object data models, and create a mechanism for greater sensitivity in modeling spatial phenomena.

While object fields allow definition of entities in field data models, there is a commensurate need to incorporate vagueness in GIS. Humans often use vague relational terms like near, fairly close, or quite far. What those descriptions mean is entirely context-dependent. Scotland is fairly close to Wales, and my house is fairly close to the park. Michael Worboys' (2001) goal is to understand how people perceive nearness; and test formal methods of incorporating these concepts. "Vague" is used in this context to describe relations between two geographical entities that are borderline in terms of their nearness relations. The problem arises as you move farther away from a "close" place; at what point does it become far away? If a student bicycles away from the university, is she still close to it at 1 km, 2, 3, or 4? At some point, she is considered far away (in the local context), but there is swath of territory in which the near and far are indistinguishable. Worboys investigated three different formal systems that could be used to incorporate vague (indeterminate) relations of nearness between geographical objects: multivalued logics (for example, four possible values instead of two), fuzzy logic, and nearness neighborhood boundaries. Each offered some advantage in terms of representation of vagueness in geographical relations, and extends the links between conceptual and formal ontologies.

Worboys' research follows a related trajectory: using cognitive studies to understand how people conceptualize and express spatial relations. In 1998, Shariff et al. asked subjects to describe multiple representations of spatial relationships between a road and a pond based on drawings. What the study revealed was that humans tend to describe the relationship between spatial entities in relative terms. Topology or relationships based on contiguity and adjacency are far more important to people than metrics. Topology refers to relationships like North Dakota is above South Dakota, or Norway is west of Sweden. GISystems, unlike human perception relies on geometry to describe spatial relations. There is clearly a disjuncture between the metric focus of GISystems and human spatial cognition. Shariff's and Egenhofer subsequent paper (1998) described a way to use set theory to formalize up to 512 different conceptual relationships between a lake and a pond. In normal circumstances, only 19 relationships were pertinent, but topology can be complicated. Lakes often contain islands, for instance, and islands can contain lakes. Roads and ponds can have complex concave and convex shapes that significantly expand the number of possible relationships between the two. This research was important for two reasons: it enforced the value of incorporating cognition (an expression of the encompassing *ontology*) over geometry in GIScience, while also offering a formal mechanism for incorporating the insight.

New research on cognition, affordances, naïve spatial categories, and vagueness relations each point to the shortcomings of present GISystems in representing the breadth and extent of spatial relations. Their incorporation into the GIScience research agenda points, however, to possibilities for incorporation of a wider range of ontologies into GIS. This has been a compressed and necessarily restricted discussion of ontology research in GIScience. There are, however, numerous researchers and approaches to the ontology problem. This brief synopsis does afford a glimpse of the complex issues at the intersection of cognition, classification, formalization, and their relationship to data models in the digital realm. The pertinent lesson is that GIScientists are not confined to technical solutions. Rather there is a significant movement to dimensionalize data and representation in digital environments – one that takes different epistemologies (which result in varying ontologies) into account. Any previous assumption that data models can convey the full range of human perception or expression has been cast aside in favor of developing means of digital implementation that explicitly identify epistemology and ontology. Ironically the objective of ontology research to extend the basis and extent of representation in GIS is closely related to the goals of feminism and GIS researchers.

Feminism and GIS: Blurring the Boundaries of the Discipline

GIS is associated in the minds of many – and often unfairly – with a technical realm of geography that is dominated by "geeks" (often men) and computers. It has been considered very distant from feminist geography, and human geography in general (see Chapter 2). This distance is not intrinsic to GIS, but the result of different cultures and practices of geographical inquiry. Stacy Warren (2003) points out that there is nothing inherently masculine about GIS. Every element of the present technology and its implementation is subject to change. Moreover, the term feminism is itself inclusive; social scientists frequently use it to refer generally to concerns about equality, marginalization, distribution of wealth, social justice, and allocation of power. Men can be feminists; feminism is not the exclusive domain of women. Feminist researchers familiar with GIS share a commitment to the use of GIS to illuminate spatial dimensions of social and physical phenomena, but, like all GIScience researchers, they are also concerned with the limitations of GIS.

The most traditional instance of GIS and gender uses GIS to study spatial problems that involve women and other disenfranchised communities. This research direction has the potential to influence people's lives in practical ways, given the close relationship between GIS maps and public policy. In one example, Sarah Elwood (2000) studied the use of GIS by a community group to lobby several levels of government for more resources. The community was somewhat downtrodden, but eager to obtain public funds to rehabilitate its physical infrastructure. The Powderhorn Park Neighborhood Association (PPNA) adopted GIS and database technologies to reframe representations of their community to municipal and state governments. Elwood was able to illustrate that adoption of GIS had a marked effect on the way that the PPNA interacted with various levels of government. Geographic technologies shifted the power dynamics between the state and community. For example, the PPNA was initially excluded from a state rehabilitation fund in which qualification was determined by evidence of significant improvement in housing conditions. They successfully challenged this ruling using a combination of GIS and anecdotal evidence. Elwood's study illustrated that geographic technologies led to an increased legitimacy for this association, and therefore a redistribution of social and political power. She also determined that use of geographical technologies led to concurrent reliance on rational planning procedures that may create barriers to participation for some residents, not surprisingly those with the least social power.

Sara McLafferty (2002) provides a very different example of feminist concerns intersecting with GIS. She offers the story of a group of women from Long Island who had coalesced to analyze high rates of breast cancer in their community. They wanted to analyze whether environmental problems were linked to the incidence of disease. Initially the group petitioned the State of New York to investigate, but the state scientist focused on explanations that focused on individual behavior - such as high fat consumption. They subsequently contacted McLafferty's research group at Hunter College and asked them to investigate questions that might link breast cancer clusters to cul-de-sacs where sewers merged, golf course runoff, nuclear pollution from a nearby plant, pesticides, or contaminated soil. This initial investigation was accompanied by political pressure by the residents of Long Island – to which a federal senator from New York State responded. The result was a law requiring the National Cancer Institute to examine possible environmental hazards related to high rates of breast cancer on Long Island using GIS. Preliminary, inconclusive results are available from The Long Island Breast Cancer Study Project (*http://www.health-li.com/*).

The paradigm of proof in epidemiology is based on traditional science which dictates that a significant percentage of incidence must be positively correlated with causes. In this case, the study is stalled partly because of difficulties inherent to epidemiology. Devra Davis (2002), a toxicologist, notes that historical data of persons and environment are difficult to collate, and early life exposures are often more pertinent than where a women is living when she is diagnosed with breast cancer. There is a long latency period associated with breast cancer, and making firm connections between cause and effect are extremely difficult. Though each of the environmental agents that the Long Island women suspect to be linked to their cancer is under suspicion, there is no hard statistical evidence. One of the difficulties epidemiologists face is the rigor to which statistical connections are subject. For every positive association, there is a critic who can undermine the numbers. The trouble is that people are mobile, memories unreliable, and cancer is caused by a host of overlapping factors. Moreover, for every positive association, there is a critic who can remanipulate the numbers to "disprove" the suggested correlation – and often provide polluters with arguments for delaying or ending regulation. GIS is one way to investigate data in a more qualitative manner.

McLafferty illustrates the extent to which GIS can be used to express exploratory queries, as well as include context and a "sense" of place. Long Island women wanted to include oral narratives which described past environmental practices, sketches of environmental features, and historical photographs. One of the more interesting aspects of this story, however, is the extent to which the Long Island women were successful in affecting policy to the extent that a geographically specific law was enacted. It underscores the persuasive powers of GIS based on its visuality – and authoritative presentation – without the constraints of "hard" numerical data. It offers a method of expressing suspected relationships between myriad factors. Moreover, a simple map with only one variable can powerfully express spatial relationships for which there is no explanation. For instance, the map of US breast cancer mortality illustrates that women in Northeastern states and states surrounded the Great Lakes are afflicted with breast cancer at a rate of close to 100 women per 100,000 as illustrated in Figure 5.3. In the rest of the country, only small pockets of California and Washington have rates this high, except for the state of Nevada where the entire state is colored black to represent the same high rate. Nevada was the site of sustained nuclear testing in the 1950s which likely explains (but does not prove) this connection.

Another research direction consistent with feminist GIS is using visualization to integrate qualitative material to express social relations in different ways. Mei-Po Kwan is at the forefront of this shift, and could be said to have initiated the feminism and GIS discussion. Kwan's superior technical and analytical abilities combined with her understanding of qualitative versus quantitative debates in geography have enabled her to experiment with new forms of the technology. Her earlier work (1998; 1999) focused on using network-based GIS to portray space–time relations of women (as described in Chapter 2). Surveys of individual women were used in conjunction with rigorous quantitative techniques to illustrate that women frequently have reduced access to urban events and amenities compared to men. Indeed, little relationship between travel time and extracurricular activity was found, primarily because women are subject to numerous time demands. The spatial "fixity" of women was portrayed using a number of GIS and statistical techniques, including 3D GIS, to illustrate activity patterns over time and space.

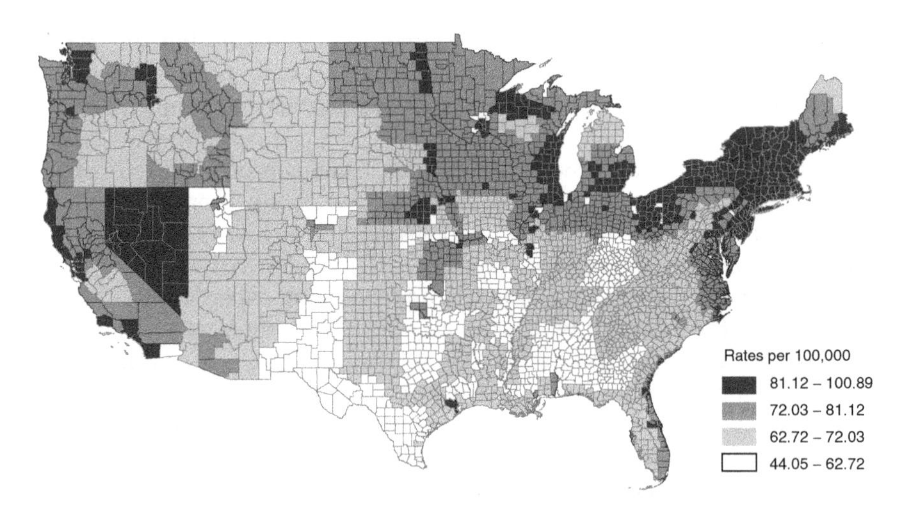

Figure 5.3 *Map illustrating breast cancer distribution by county in the United States.*
The borders between some counties have been generalized. Note the high concentration in Nevada – likely linked to nuclear testing in the 1950s and 1960s.

Source: *This figure is adapted from Davis (2002).*

The results inform her subsequent investigation into the problems and potentials of extending a feminist politic using GIS visualization. This body of work asks the question: "is GIS compatible with alternate epistemologies associated with feminism, including a greater reliance on quantitative material?" Kwan's tentative answer is yes, with the qualification being that off-the-shelf software is not presently equipped to accommodate qualitative information in analysis, though she demonstrates the viability of its inclusion (Kwan, 2002).

Ontologies and feminist politics are two sites of inquiry for GIScience research. Their concern with representation help to make GIS more inclusive, more nuanced, and more useful for representing the complexities of geography and spatial relations. The fields are linked by emphasis on inclusion and more nuanced representation of a broader range of entities. Ontology research and feminist GIS have the potential to move the emphasis in GIS from representing fixed spatial entities like roads, bridges, farms, electrical utilities, and such to offering multiple vantage points depending on epistemology, and facilitating exploratory research between multiple variables that are difficult to incorporate into statistical formulae. In combination, these research sites, and the GIScience research community that supports them are poised to develop an enhanced representational framework that surpasses the technical parameters and epistemological weaknesses identified by human geographers in the 1990s.

Conclusion: Demarcating a Fuzzy Territory Between Systems and Science

This book has initiated exploration of the intellectual territory that constitutes GIS. It has also delved into a number of practical issues that affect the practice of GIS from data to spatial analysis. Many of these issues and applications are central to both GISystems and GIScience. Indeed the two interpretations of the acronym GIS are closely related. There would be no GIScience without the systems, and, conversely, the science drives new developments in software. This relationship is synergestic with the growing rise in systems users contributing to the need for new ways of envisioning spatial relations. Certainly the number of GIS users is increasing. ESRI, one of the leading developers of GIS software sells almost $500,000 (USD) of software annually. Combined with Microsoft's entry into the GIS world with MapPoint software, the number of users is poised to rise dramatically. It is estimated that this growth will remain unconstrained for the foreseeable future as the market is nowhere near saturated. One could argue that the potential for GIS users is inherently limited. As a counterpoint, few computer users used or saw the need for a spreadsheet before Excel was introduced as part of MicroSoft Office® software suite. Today, the use of spreadsheets is pervasive; they have become standard office equipment. It is likely that GIS use will continue to grow on a similar trajectory. As this happens, spatial awareness among lay users will rise dramatically. In 1997, the United States National Research Council published a book arguing that American science was at a disadvantage because it had largely ignored spatial dimensions of many phenomena. The rise in GIS use is an effective, if unexpected, counter to this trend. New GIS users will be at a disadvantage, however, if they presume that by virtue of having the requisite software, they are in a position to embark on spatial analysis of various sundry phenomena for which data exist. It is at this juncture that the demarcation between science and systems becomes blurred.

GIS instructors are constantly at pains to remind students that the maps they print are not definitive statements about spatial relations, but representations of a particular point of view. In the absence of educated users, GIS becomes a tool to make map-pictures rather than represent or predict spatial phenomena. Moreover, the usefulness of any GIS map depends not only on the intellectual assumptions inherent in its generation, but on the extent to which it conforms to standards of practice developed in GIScience. Such practices are the basis for legitimation of the results. This book has attempted to explicate some of the dangers of adopting GISystems with little understanding of the spatial science or intellectual territory that informs them. An abbreviated list of cautions begins with the specificity of data models.

The GIS realm of present possibilities is divided into fields and objects with raster and vector mechanisms or structures for displaying either. In practice, however, raster (or grid cell) representation is usually reserved for the field model, and vector is used to describe either fields or objects. The ontological consequences of each method of representation is different with field data models implemented in raster requiring the least ontological commitment. Transitions in the characteristics of a neighborhood or community are easily recorded in nuanced fashion, thereby avoiding the use of fixed categories that assume internal homogeneity across a large area. However, vector representations of geographical objects such as bridges, roads, farms, and cities are thought to more closely mimic human cognitive categories, and their use conforms to well-defined social and intellectual traditions of describing our surroundings based on entities rather than referring to vague clusters of similar (but often not identical) cell attributes. Despite emerging research that transcends the dualism of fields and objects, they remain the basis for representation. Object fields and ontology editors for GIS do not dispose of data models; they work with them. Users still need to understand the ontological implications of choosing raster or vector.

The quality of the data that populate data models constitute the best indicator of the quality of the resulting spatial analysis. Poor quality or inappropriate data can invalidate the results of an analysis, despite a very persuasive-looking cartographic product. As GIS use increases, queries about the integrity of analysis are extending from the map to the data and the models used to manipulate them. In the past, a GIS map presented at a town meeting or land use hearing was often sufficient to persuade the audience that a given outcome would have a calibrated effect as exemplified in the evidence (map). The burden of proof has, however, extended to every facet of GIS production. This reflects a cultural/scientific insistence on "objectivity" however vague the term may be to some. A demand for scientific "objectivity" underlies the political and legal basis of our society. "The whole point of GIS' popularity in a lot of agencies has always been that it... allows people to say exactly what they've done and defend it in court" (Michael Goodchild, 1998, personal interview). As researchers and users alike have found that the results of GIS vary depending on the quality of data, the demand for justifying choice and integrity of data has intensified. In this GIS follows statistics which have long been subject to intense scrutiny in legal situations.

In the past two decades, the scope of GIS has extended greatly. The ability to perform complex analyses that yield predictions and assist in spatial decision making have aided in this transition, but it has also been boosted by a greater availability of spatial data from multiple sources. The integration of such data frequently permits analysts to understand complex processes that cover wide swaths of territory. The combination of these two influences has aided in the migration of GIS from a technique that was largely associated with in-house proprietary projects like recording property lines for cadastral management and managing public lands to understanding factors that influence global warming or identifying communities at risk for increased mortality. The difficulty of integrating large, multisource data sets must be measured against the great benefit of insight into multifaceted social and physical processes. Increased awareness of the user community combined with more consistent use of metadata are keys to data integration that can withstand scrutiny.

Better data practices are integral to reliable GIS results, but they cannot compensate for poorly conceived spatial analysis that is inappropriate to the problem at hand. Most GIS software programs are wonderfully robust. The user can dictate a spatial join between data that should not be merged, and the software will execute the command without a whimper. If the resulting data set is then subject to further manipulation, it may be difficult to detect early errors that contravene logic, even for an expert. There is no better insurance against inappropriate and potentially dangerous wrong analyses than solid training in GIScience. The importance of GIS education and awareness extend not only to technical users, but to potential researchers and supervisors who may never touch the keyboard, but are responsible for studying or managing phenomena using GIS. This text constitutes a short introduction to the potential for applied GIS and the underlying intellectual territory that validates its use. The line between the two is fuzzy, and should remain so in the interests of a better, more reliable, socially responsible GIS.

References

References to Chapter 1

BangaloreIT. What is e-governance? http://www.bangaloreit.com/html/egovern/egovern.htm. Accessed January 29, 2003.

Burrough, Peter A., and Andrew U. Frank, eds 1996. *Geographic Objects with Indeterminate Boundaries*. London: Taylor & Francis.

Campari, I. 1996. Uncertain Boundaries in Urban Space. In *Geographic Objects with Indeterminate Boundaries*, ed. by P.A. Burrough and A.U. Frank. Bristol, PA: Taylor & Francis, 57–69.

Chrisman, N. 1998. Academic Origins of GIS. In *The History of Geographic Information Systems*, ed. by T. Foresman. W. Upper Saddle River, NJ: Prentice Hall, 33–43.

Chrisman, N. 1997. *Exploring Geographic Information Systems*. New York: John Wiley & Sons, Inc.

Chrisman, N. 1988. The Risks of Software Innovation: A Case Study of the Harvard Lab. *The American Cartographer* 15 (3):291–9.

Daratech. *Geographic Information Systems: Markets & Opportunities*. Daratech 2002. [cited June 21, 2002.] Available from http://www.daratech.com/gis/markets_&_opportunities.shtml.

ESRI. GIS touches our everyday lives. http://www.esri.com/company/gis_touches/start.html. Accessed January 23, 2003.

Fisher, Peter F., and Jo Wood. 1998. What is a Mountain? Or The Englishman who went up a Boolean Geographical Concept but Realised it was Fuzzy. *Geography* 83 (3):247–56.

Flowerdew, Robin, and James Pearce. 2001. Linking Point and Area Data to Model Primary School Performance Indicators. *Geographical and Environmental Modeling* 5 (1):23–41.

Foresman, T., ed. 1998. *The History of Geographic Information Systems : Perspectives from the Pioneers*. Upper Saddle River, NJ: Prentice Hall.

Goodchild, M.F. 1992. Geographical Information Science. *International Journal of Geographical Information Systems* 6 (1):31–45.

Goodchild, M.F. 1995. Geographic Systems Information and Research. In *Ground Truth*, ed, by J. Pickles. New York: Guildford Press: 1–30.

Gregory, D. 1994. Ontology. In *The Dictionary of Human Geography*, ed. by R. J. Johnston, D. Gregory, and D. M. Smith. Oxford: Blackwell Publishing: 426–9.

Koch, Tom. 2003. The Map as Intent: Variations on the theme of John Snow. Unpublished manuscript.

Latour, B. 1987. *Science in Action*. Cambridge MA, Harvard University Press.

Poster, M. 1996. Databases as Discourse, or Electronic Interpellations. In *Computers, Surveillance and Privacy*, ed. by D. Lyon and E. Zureik. Minneapolis: University of Minnesota Press, 175–92.

Philo, C., R. Mitchell, and A. More. 1998. Guest Editorial: Reconsidering Quantitative Geography: The Things that Count. *Environment and Planning A* 30 (2):191–202.

Rhind, D.W. 1988. Personality as a Factor in the Development of a New Discipline: The Case of Computer-Assisted Cartography. *The American Cartographer* 15 (3):277–89.

Schuurman, N. 1999a. Speaking With the Enemy? An Interview With Michael Goodchild. *Environment and planning: D Society and Space* 17 (1):1–15.

Schuurman, N. 1999b. Critical GIS: Theorizing an Emerging Discipline. *Cartographica* 36 (4): 1–109

Schuurman, N., and G. Pratt. 2002. Care of the Subject: Feminism and Critiques of GIS. *Gender, Place, and Culture* 9 (3):291–9.

Smith, B., and Mark, D.M. 2001. Geographical Categories: An Ontological Investigation. *International Journal of Geographical Information Science* 15 (7):591–612.

Tomlinson, R.F. 1989. Presidential Address: Geographic Information Systems and Geographers in the 1990s. *The Canadian Geographer* 33 (4):290–8.

Tomlinson, R.F. 1988. The Impact of the Transition From Analogue to Digital Cartographic Representation. *The American Cartographer* 15 (3):249–61.

Tufte, E.R. 1997. *Visual Explanations: Images and Quantities, Evidence and Narrative*. Cheshire, CT: Graphics Press.

Unwin, D. 2001. *GIS and the Peopling of an Industry*. Paper read at GIS Workshop on *A Changing Society*, May 17–20, 2001, at The Ohio State University.

Warren, S. 2003. The Utopian Potential of GIS. *Cartographica* forthcoming.

Watson, J. 1969. *The Double Helix*. New York: Mentor.

References to Chapter 2

Baker, V.R. 2000. Conversing with the Earth: The Geological Approach to Understanding. In *Earth Matters: The Earth Sciences, Philosophy, and the Claims of Community*. R. Frodeman. Upper Saddle River, NJ, Prentice Hall: 2–10.

Berry, J. 1999. Is Technology Ahead of Science? *GeoWorld* 12 (2): 28–9.

Brassel, K.E. and R. Weibel, 1988. A review and conceptual framework of automated map generalization. *International Journal of Geographical Information Systems* 2(3): 229–44.

Burrough, P. A. 1996. Natural Objects with Indeterminate Boundaries. *Geographic Objects with Indeterminate Boundaries.* P.A. Burrough and A.U. Frank. Bristol, PA,:Taylor & Francis, 3–28.

Buttenfield, B. P. and R. P. McMaster, 1991. *Map Generalization: Making Rules for Knowledge Representation.* New York, Wiley.

Couclelis, H. 1992. People Manipulate Objects but Cultivate Fields: Beyond the Raster – Vector Debate in GIS. In *Theories and Methods of Spatial-Temporal Reasoning in Geographic Space* ed. by A. U. Frank, I. Campari and U. Formentini. Berlin: Springer-Verlag, 65–77.

Couclelis, H. 1999. Space, Time, Geography. In *Geographical Information Systems: Principles, Techniques, Management and Applications,* ed. by P. A. Longley, M. F. Goodchild, D. J. Maguire and D. W. Rhind. New York: John Wiley & Sons, 29–38.

Curry, M. 1997. The Digital Individual and the Private Realm. *Annals of the Association of American Geographers* 87 (4): 681–99.

Demeritt, D. 2001. The Construction of Global Warming and the Politics of Science. *Annals of the Association of American Geographers* 91 (2): 307–37.

Dutton, G. 1977. *Proceedings of the First Internatinoal Study Symposium on Topological Data Structures for Geographical Information Systems.* First International Study Symposium on Topological Data Structures for Geographical Information Systems, Cambridge, MA., Laboratory for Computer Graphics and Spatial Analysis.

Dyson, F. 1999. *The Sun, the Genome, and the Internet: Tools of Scientific Revolutions.* New York: Oxford University Press.

Frank, A.U. 2001. Tiers of Ontology and Consistency Constraints in Geographical Information Systems. *International Journal of Geographical Information Science* 15 (7): 667–78.

Goodchild, M.F. 1991. Just the Facts. *Political Geography Quarterly* 10: 192–3.

Goodchild, M.F. 1992. Geographical Information Science. *International Journal of Geographical Information Systems* 6 (1): 31–45.

Goodchild, M.F. 1995. Geographic Systems Information and Research. In *Ground Truth* ed. by J. Pickles. New York: Guildford Press, 31–50.

Goss, J. 1995. Marketing the New Marketing: The Strategic Discourse of Geodemographic Information Systems. In *Ground Truth* ed. by J. Pickles. New York: Guildford Press, 130–70.

Gregory, D. 1978. *Ideology, Science and Human Geography.* New York: St. Martin's Press.

Gregory, D. 1994a. Ontology. In *The Dictionary of Human Geography* ed. by R.J. Johnston, D. Gregory and D.M. Smith. Oxford: Blackwell, 426–9.

Gregory, D. 1994b. Positivism. In *The Dictionary of Human Geography* ed. by R.J. Johnston, D. Gregory and D.M. Smith. Oxford: Blackwell, 455–7.

Gregory, D. 1994c. Pragmatism. In *The Dictionary of Human Geography* ed. by R.J. Johnston, D. Gregory and D.M. Smith. Oxford: Blackwell, 471–2.

Gregory, D. 1994d. Realism. In *The Dictionary of Human Geography* ed. by R.J. Johnston, D. Gregory and D.M. Smith. Oxford: Blackwell, 500.

Gregory, D. 2000. Realism. In *The Dictionary of Human Geography*, 2nd edn, ed. by R.J. Johnston, D. Gregory, G. Pratt and M. Watts. Oxford: Blackwell, 673–6.

Gruber, T. 1995. Toward Principles for the Design of Ontologies used for Knowledge Sharing. *International Journal of Human-Computer Studies* 43: 907–8.

Haraway, D. 1991. *Simian, Cyborgs and Women: The Reinvention of Nature*. New York: Routledge.

Haraway, D. 1997. enlightenment@science wars.com: A Personal Reflection on Love and War. *Social Text* 50: 123–9.

Harley, B. 1989. Deconstructing the Map. *Cartographica* 26: 1–20.

Harris, T., D. Weiner, et al. 1995. Pursuing Social Goals Through Participatory GIS: Redressing South Africa's Historical Political Ecology. In *Ground Truth* ed. by J. Pickles. New York: Guildford Press, 196–222.

Harvey, F. and N. R. Chrisman, 1998. Boundary Objects and the Social Construction of GIS Technology. *Environment and Planning A* 30: 1683–94.

Harvey, F. 2003. The Linguistic Trading Zones of Semantic Interoperability. In *Representing GIS* ed. by D. Unwin. London: John Wiley & Sons, Inc. forthcoming.

Heidegger, M. 1982. *The Question Concerning Technology and Other Essays*. New York: Harper Collins.

Jordan, T.G. 1988. The Intellectual Core: President's Column. *AAG Newsletter* 23: 5. Kuhn, T. 1970. *The Structure of Scientific Revolutions*. Chicago: University of Chicago Press.

Kuhn, W. 1994. *Defining Semantics for Spatial Data Transfers*. Sixth International Symposium on Spatial Data Handling. Edinburgh: Taylor & Francis.

Kwan, M. 1998. Space-Time and Integral Measures of Individual Accessibility: A Comparative Analysis Using a Point-based Framework. *Geographical Analysis* 30 (3): 191–16.

Kwan, M. 1999. Gender, the Home–Work Link, and Space–Time Patterns of Nonemployment Activities. *Economic Geography* 75 (4): 370–94.

Latour, B. 1987. *Science in Action*. Cambridge, MA: Harvard University Press.

Latour, B. 1988. *The Pasteurization of France*. Cambridge, MA: Harvard University Press.

Latour, B. 1999. On Recalling ANT. In *Actor Network Theory and After* ed. by J. Law and J. Hassard. Oxford: Blackwell Publishers/The Sociological Review, 25.

Law, J. 1994. *Organizing Modernity*. Oxford: Blackwell.

MacEachren, A.M. 1994. Visualization in Modern Cartography: Setting the Agenda. In *Visualization in Modern Cartography* ed. by A.M. MacEachren and D.R.F. Taylor. Tarrytown, NY: Elsevier, 1–12.

MacEachren, A.M., M. Wachowicz, et al. 1999. Constructing Knowledge from Multivariate Spatiotemporal Data: Integrating Geographical Visualization with Knowledge Discovery in Databases. *International Journal Of Geographical Information Systems* 13 (4): 311–34.

Mark, D.M. 1993. Towards a Theoretical Framework for Geographic Entity Types. In *Spatial Information Theory: A Theoretical Basis for GIS* ed. by A. U. Frank and I. Campari. Berlin: Springer-Verlag, 270–83.

Mark, D.M. 1999. Spatial Representation: A Cognitive View. In *Geographical Information Systems: Principles, Techniques, Management and Applications* ed. by P.A. Longley, M.F. Goodchild, D.J. Maguire, and D.W. Rhind. New York: Wiley, 81–9.

Mercer, D. 1984. Unmasking Technocratic Geography. In *Recollections of a Revolution* ed. by M. Billinge, D. Gregory, and R. Martin. London: Macmillan.

Monmonier, M. 1996. *How to Lie With Maps*. Chicago: University of Chicago Press.

Openshaw, S. 1991. A View on the GIS Crisis in Geography, Or Using GIS to Put Humpty-Dumpty Back Together Again. *Environment and Planning A* 23 (5): 621–8.

Openshaw, S. 1992. Further Thoughts on Geography and GIS: A Reply. *Environment and Planning A* 24: 463–6.

O'Tuathail, G. 1996. *Critical Geopolitics: The Politics of Writing Global Space*. Minneapolis: University of Minneapolis Press.

Pickering, A. 1995. *The Mangle of Practice: Time, Agency, & Science*. Chicago; University of Chicago Press.

Pickles, J. 1995. Representations in an Electronic Age: Geography, GIS, and Democracy. In *Ground Truth* ed. by J. Pickles. New York: Guildford Press, 1–30.

Pickles, J. 1997. Tool or Science? GIS, Technoscience, and the Theoretical Turn. *Annals of the Association of American Geographers* 87: 363–72.

Rouse, J. 1996. *Engaging Science. How to Understand Its Practices Philosophically*. Ithaca: Cornell University Press.

Raper, J. 1999. Spatial Representation: The Scientist's Perspective. In *Geographical Information Systems: Principles, Techniques, Management and Applications* ed. by P.A. Longley, M.F. Goodchild, D.J. Maguire, and D.W. Rhind. New York: Wiley, 71–80.

Raper, J. 2000. *Multidimensional Geographic Information Science*. New York: Taylor & Francis.

Roberts, S.M. and R.H. Schein, 1995. Earth Shattering: Global Imagery and GIS. In *Ground Truth* ed. by J. Pickles. New York: Guildford Press, 171–95.

Schuurman, N. 1999a. Speaking With the Enemy? An Interview With Michael Goodchild. *Environment and Planning: D Society and Space* 17 (1): 1–15.

Schuurman, N. 1999b. Critical GIS: Theorizing an Emerging Discipline. *Cartographica* 36 (4): 1–109.

Schuurman, N. 2000. Trouble in the Heartland: GIS and its critics in the 1990s. *Progress in Human Geography* 24 (4): 569–590.

Schuurman, N. 2002. Reconciling Social Constructivism and Realism in GIS. *ACME: An International E-Journal for Critical Geographies* 1 (1): 75–90.

Schuurman, N. and G. Pratt, 2002. Care of the Subject: Feminism and Critiques of GIS. *Gender, Place and Culture* 9 (3): 291–9.

Sheppard, E. 1995. GIS and Society: Toward a Research Agenda. *Cartography and Geographic Information Systems* 22 (1): 5–16.

Sieber, R. 2003. Public Participation Geographic Information Systems Across Borders. *The Canadian Geographer* 47 (1): 50–61.

Sismondo, S. 1996. *Science without Myth: On Constructions, Reality and Social Knowledge*. Albany: State University of New York Press.

Smith, B. and D.M. Mark, 1998. Ontology and Geographic Kinds. *Proceedings from the 8th International Symposium on Spatial Data Handling*, 308–20.

Smith, B. and D.M. Mark, 2001. Geographical Categories: An Ontological Investigation. *International Journal of Geographical Information Science* 15 (7): 591–612.

Smith, N. 1992. History and Philosophy of Geography: Real Wars, Theory Wars. *Progress in Human Geography* 16: 257–71.

Spivak, G.C. 1987. Can the Subaltern Speak? In *Marxism and the Interpretation of Culture* ed. by C. Nelson and L. Grossberg. New York/London: Routledge.

Taylor, P. J. 1990. GKS. *Political Geography Quarterly* 9: 211–12.

Taylor, P.J. and R.. Johnston 1995. GIS and Geography. In *Ground Truth* ed. by J. Pickles. New York: Guildford Press: 68–87.

Tomlinson, R.F. 1984. *Panel Discussion: Technology Alternatives and Technology.* Computer Assisted Cartography and Geographic Information Processing: Hope and Realism, Ottawa, Canadian Cartographic Association.

Watson, J. 1969. *The Double Helix*. New York: Mentor.

Weiner, D., T. Warner, et al. 1995. Apartheid Representations in a Digital Landscape: GIS, Remote Sensing and Local Knowledge in Kiepersol, South Africa. *Cartography and Geographic Information Systems* 22: 30–44.

Winter, S. 2001. Ontology: Buzzword or Paradigm Shift in GI Science? *International Journal of Geographical Information Science* 15 (7): 587–90.

Ziauddin, S. 2000. *Thomas Kuhn and the Science Wars*. New York: Totem Books.

References to Chapter 3

Bowker, G. 2000. Mapping Biodiversity. *International Journal of Geographical Information Science* 14 (8): 739–54.

Bowker, G., and Star, S. L. 2000. *Sorting Things Out: Classification and its Consequences*. Cambridge, MA: MIT Press.

Burrough, P.A., and Mcdonnell, R. 1998. *Principles of Geographic Information Systems*. Oxford: Oxford University Press.

Desbarats, A.J., M.J. Hinton, et al. 2001. Geostatistical Mapping of Leakance in A Regional Aquitard, Oak Ridges Moraine Area, Ontario, Canada. *Hydology Journal* 9: 79–96.

Devogele, T., Parent, C., and Spaccapietra, S. 1998. On Spatial Database Integration. *International Journal of Geographical Information Science* 12 (4): 335–52.

Goodchild, M., F. and Proctor, J. 1997. Scale in a Digital Geographic World. *Geographical and Environmental Modelling* 1 (1): 5–23.

Goodchild, M.F., Egenhofer, M., and Fegeas, R. 1997. Interoperating GISs: Report of a Specialist Meeting Held Under the Auspices of the Varenius Project.

Gregory, D. 1994. Pragmatism. In *The Dictionary of Human Geography*. ed. by R.J. Johnston, D. Gregory and D.M. Smith. Oxford, Blackwell: 471–2.

Kashyap, V., and Sheth, A. 1996. Semantic and Schematic Similarities between Database Objects: A Context-Based Approach. *The VLDB Journal* 5: 276–304.

Kraak, M.J., and Ormeling, F.J. 1996. *Cartography: Visualization of Spatial Data*. Toronto: Prentice Hall.

Laurini, R. 1998. Spatial Multi-Database Topological Continuity and Indexing: A Step Towards Seamless GIS Data Interoperability. *Geographical Information Science* 12 (4): 373–402.

Logan, C., H.A.J. Russell, et al. 2001. Regional Three-Dimensional Stratigraphic Modelling of the Oak Ridges Moraine Areas, Southern Ontario. *Geological Survey of Canada Current Research* 2001-D1.

Mark, D.M. 1993. Towards a Theoretical Framework for Geographic Entity Types. In *Spatial Information Theory: A Theoretical Basis for GIS* ed. by A.U. Frank and I. Campari. Berlin: Springer-Verlag, 270–83.

National Desk, 2001. *U.S. Will Not Adjust 2000 Census Figures*, Wednesday, March 7.

O'Connor, D.R. 2002. Report of the Walkerton Inquiry: The Events of May 2000 and related issues. Toronto, Ontario Ministry of the Attorney General. www.walkertoninquiry.com/report1/index.html#summary. Accessed Nov. 8, 2002.

Office of Research and Statistics, 2002. www.ors.state.sc.us/digital/census.asp. Accessed September 4, 2002.

Pima County Association of Governments, 2002.

Russell, H.A.J., C. Logan, et al. 1996. Geological Investigations: Subsurface Data, Geological Survey of Canada. 2001.

Schuurman, N. 1999. Critical GIS: Theorizing an Emerging Discipline. *Cartographica* Monograph 36 (4): 1–109.

Schuurman, N. 2002. Reconciling Social Constructivism and Realism in GIS. *ACME: An International E-Journal for Critical Geographies* 1 (1): 75–90.

http: // www.pagnet.org/Population/census/Default.htm. Accessed September 2, 2002.

References to Chapter 4

Adriaans, P., and D. Zantinge. 1996. *Data Mining*. New York: Addison-Wesley.

Atkinson, P. M. and N. Tate, J. 2000. Spatial Scale Problems and Geostatistical Solutions: A Review. *The Professional Geographer* 52 (4): 607–23.

Bernhardsen, Tor. 1999. *Geographic Information Systems: An Introduction*. Toronto: John Wiley & Sons, Inc.

Blom, T., and R. Savolainen-Mdntyjdrvi. 2001. *GIS and Health*. 2001 [cited June 13 2001]. Available from www.shef.ac.uk/uni/academic/D-H/gis/healrepo. html.

Chen, F., and J. Delaney. 1998. Expert Knowledge Acquisition: A Methodology for GIS Assisted Industrial Land Suitability Assessment. *Urban Policy and Research* 16 (4): 301–15.

Chen, F., and J. Delaney. 1999. Integrating GIS and Environmental Pollution Modeling for Industrial Land-use Planning. Paper read at Thirteenth Annual Conference on Geographic Information Systems, at Vancouver, BC.

Chrisman, N. 1997. *Exploring Geographic Information Systems*. New York: John Wiley & Sons, Inc.

Cloud, J. and K. Clark. June 20–2, 1999. *The Fubini Hypothesis: The Other History of Geographic Information Science*. Geographic Information and Society, University of Minnesota.

Crampton, Jeremy. 2002. Guest editorial. *Environment and planning D: Society and Space* 20: 631–5

Csillag, Ferenc, Marie-Josée Fortin, and Jennifer L. Dungan. 2000. On the Limits and Extensions of the Definition of Scale. *Bulletin of the Ecological Society of America* 81 (3): 230–2.

Curry, M. 1997. The Digital Individual and the Private Realm. *Annals of the Association of American Geographers* 87 (4): 681–99.

Curry, M. 1998. *Digital Places: Living With Geographic Information Technologies*. New York: Routledge.

DeMers, M.N. 2000. *Fundamentals of Geographic Information Systems.* 2nd edn Toronto: Wiley & Sons, Inc.

Dobson, J. 1993. Automated Geography. *Professional Geographer* 35: 135–43.

Dragicevic, S., N. Schuurman, and M. Fitzgerald. 2003. The Utility of Exploratory Spatial Data Analysis in the Study of Tuberculosis Incidences in an Urban Canadian Population. *Cartographica*: forthcoming.

Dunn, J.R. and M.V. Hayes. 2000. Social Inequality, Population Health, and Housing: A Study of Two Vancouver Neighborhoods. *Social Science and Medicine* 51: 563–87.

Dye, C., Williams, B. G., Espinal, M., A., and Raviglione, M. C. 2002. Erasing the World's Slow Stain: Strategies to beat Multidrug-resistant Tuberculosis. *Science* 295: 2042–2046

The Economist. 2003. The Best Thing Since the Bar-Code. *The Economist* February 8, 2003: 57–8.

Floyd, K., Blanc, L., Raviglione, M.C., and Lee, J.-W. 2002. Resources Required for Global Tuberculosis Control. *Science* 295: 2020–41.

Foucault, M. 1979. *Discipline and Punish.* New York, Vintage Books.

Frye, N. 1982. *Divisions on a Ground: Essays on Canadian Culture.* Toronto, Anansi.

Gahegan, M. 1999. Characterizing the Semantic Content of Geographic Data, Models, and Systems. In *Interoperating Geographic Information Systems,* ed. by M. Goodchild, M. Egenhofer, R. Fegeas, and C. Kottman. Boston: Kluwer Academic Publishers.

GVRD. 2003. *Garbage and Recycling – An Introduction.* GVRD, September 12, 2002 2002 [cited February 10, 2003 2003]. Available from http: || www.gvrd.bc.ca/ services/garbage/index.html.

Harley, B. 1989. Deconstructing the Map. *Cartographica* 26: 1–20.

Heywood, I., S. Cornelius, and S. Carver. 1998. *An Introduction to Geographical Information Systems.* New York: Addison Wesley Longman.

Keylock, C.J., D.M. McClung, and M.M. Magnusson. 1999. Avalanche Risk Mapping by Simulation. *Journal of Glaciology* 45 (150): 303–14.

Longley, P.A., M.F. Goodchild, D.J. Maguire, and D.W. Rhind, eds 2001. *Geographical Information Systems and Science.* New York: John Wiley & Sons, Inc.

MacEachren, A.M. 1994. Visualization in Modern Cartography: Setting the Agenda. In *Visualization in Modern Cartography,* ed. by A.M. MacEachren and D.R.F. Taylor. Tarrytown, NY: Elsevier Science Inc.

MacEachren, A.M., M. Wachowicz, R. Edsall, D. Haug, and R. Masters. 1999. Constructing Knowledge from Multivariate Spatiotemporal Data: Integrating

Geographical Visualization with Knowledge Discovery in Databases. *International Journal of Geographical Information Systems* 13 (4): 311–34.
Martin, D. 1996. An Assessment of Surface and Zonal Models of Population. *International Journal of Geographical Information Systems* 10 (8): 973–89.
Openshaw, S. and C. Openshaw 1997. *Artificial Intelligence in Geography*. New York, John Wiley & Sons.
Poster, M. 1996. Databases as Discourse, or Electronic Interpellations. In *Computers, Surveillance and Privacy*, ed. by D. Lyon and E. Zureik. Minneapolis: University of Minnesota Press.
Schuurman, N. 1999. Critical GIS: Theorizing an Emerging Discipline. *Cartographica 36* (4): 1–109.
Schuurman, N. 2002. Reconciling Social Constructivism and Realism in GIS. *ACME: An International E-Journal for Critical Geographies* 1 (1): 75–90.
Shaw, M., D. Dorling, and R. Mitchell. 2002. *Health, Place and Society*. New York: Prentice Hall.
Shrader-Frechette, K. 2000. Reading the Riddle of Nuclear Waste: Idealized Geological Models and Positivist Epistemology. In *Earth Matters: The Earth Sciences, Philosophy, and the Claims of Community*, ed. by R. Frodeman. Upper Saddle River, NJ: Prentice Hall.
Syme, S.L. 1994. The Social Environment and Health. *Daedalus* 123 (4): 79–86.
Tate, N., J. 2000. Guest Editorial: Surfaces for GIScience. *Transactions in GIS* 4(4): 301–3.
Visvalingam, M. 1994. Visualization in GIS, Cartography and ViSc. In *Visualization in Geographical Information Systems*, ed. by H. M. Hearnshaw and D. Unwin. Toronto: John Wiley & Sons.

References to Chapter 5

Cova, T.J. and M.F. Goodchild, 2002. Extending Geographical Representation to include Fields of Spatial Objects. *International Journal of Geographical Information Science* 16 (6): 509–32.
Davis, D. 2002. *When Smoke Ran Like Water: Tales of Environmental Deception and the Battle Against Pollution*. New York: Basic Books.
Egenhofer, M.J., and D.M. Mark, 2002. *Geographic Information Science*. In *Lecture Notes in Computer Science* ed. by G. Goos, J. Hartmanis and J. van Leeuwen. Berlin: Springer-Verlag.
Elwood, S. 2000. Information for Change: The Social and Political Impacts of Geographic Information Technologies. Dissertation, Geography, University of Minnesota, Minneapolis.
Fonesca, F.T., M. J. Egenhofer, et al. 2002. Using Ontologies for Integrated Information Systems. *Transactions in GIS* 6 (3): 231–57.
Gruber, T. 1995. Toward Principles for the Design of Ontologies Used for Knowledge Sharing. *International Journal of Human-Computer Studies* 43: 907–28.
Hunter, G. 2002. Understanding Semantics and Ontologies: They're Quite Simple Really – If You Know What I Mean! *Transactions in GIS* 6 (2): 83–7.

Kuhn, Werner. 2001. Ontologies in Support of Activities in Geographical Space. *International Journal of Geographical Information Science* 15 (7): 613–31.

Kwan, M. 1998. Space–Time and Integral Measures of Individual Accessibility: A Comparative Analysis Using a Point-based Framework. *Geographical Analysis* 30 (3): 191–216.

Kwan, M. 1999. Gender and Individual Access to Urban Opportunities: A Study Using Space–Time Measures. *Professional Geographer* 51 (2): 210–27.

Kwan, M. 1999. Gender, the Home–Work Link, and Space–Time Patterns of Nonemployment Activities. *Economic Geography* 75 (4): 370–94.

Kwan, M. 2000a. Gender differences in space-time constraints. *Area* 32 (2): 145–56.

Kwan, M. 2000b. Interactive Geovisualization of Activity-Travel Patterns Using Three-Dimensional Geographical Information Systems: A Methodological Exploration with a Large Data Set. *Transportation Research Part C* 8: 185–203.

Kwan, M. 2001. Quantitative Methods and Feminist Geographic Research. In *Feminist Geography in Practice: Research and Methods*, ed. by P. Moss. Oxford: Blackwell, 160–73.

Kwan, M. 2002a. Is GIS for Women? Reflections on the Critical Discourse in the 1990s. *Gender, Place and Culture* 9 (3): 271–9.

Kwan, M. 2002b. Other GISs in Other Worlds: Feminist Visualization and Re-envisioning GIS. *Annals of the Association of American Geographers* 92 (4): 645–61.

McLafferty, S.L. 2002. Mapping Women's Worlds: Knowledge, Power and the Bounds Of GIS. *Gender, Place and Culture* 9 (3): 263–9.

National Research Council. 1997. *Rediscovering Geography: New Relevance for Science and Society*. Washington DC: National Academy Press.

Openshaw, S. 1991. A View on the GIS Crisis in Geography, Or, Using GIS to Put Humpty-Dumpty Back Together Again. *Environment and Planning* A 23: 621–8.

Smith, B., and D.M. Mark. 1998. Ontology and Geographic Kinds. In *Proceedings from the 8th International Symposium on Spatial Data Handling*, ed. by T. K. Poiker and N. Chrisman: 308–20.

Raubal, M. 2001. Ontology and Epistemology for Agent-Based Wayfinding Simulation. *International Journal of Geographical Information Science* 15 (7): 653–65.

Smith, B., and D.M. Mark, 2001. Geographical Categories: An Ontological Investigation. *International Journal of Geographical Information Science* 15 (7): 591–612.

Shariff, A. R. B. M., M. J. Egenhofer, et al. 1998. Natural-language spatial relations between linear and areal objects: the topology and mteric of English-language terms. *International Journal of Geographical Information Science* 12 (3): 215–45.

Warren, S. 2003. The Utopian Potential of GIS. *Cartographica*. forthcoming.

Winter, S. 2001. Ontology: Buzzword or Paradigm Shift in GIScience? *International Journal of Geographical Information Science* 15 (7): 587–90.

Worboys, M. F. 2001. Nearness Relations in Environmental Space. *International Journal of Geographical Information Science* 15 (7): 633–651.

Index

Note: page numbers in italics refer to illustrations or figures

CPSIA information can be obtained
at www.ICGtesting.com
Printed in the USA
LVHW08s1027110918
589789LV00012B/321/P